# SUPPLY RESPONSES IN BACKWARD AGRICULTURE

# Supply Responses in Backward Agriculture

*An Econometric Study of Chotanagpur Region*

L.N. BHAGAT

CONCEPT PUBLISHING COMPANY, NEW DELHI-110059

The Indian Council of Social Science Research gave the financial support towards the publication of this work. However, the responsibility for the contents of the book - facts, opinions, conclusions etc. - are entirely that of the author and not of the ICSSR.

ISBN 81-7022-037-8

First Published 1989



L.N. **BHAGAT** (*b*. 1948 -    )

*Published by*
Ashok Kumar Mittal
Concept Publishing Company
A/15-16, Mohan Garden Commercial Block
NEW DELHI-110059 (India)

*Composed at*
Microtech Advanced Printing Systems Pvt. Ltd.
H-13, Bali Nagar
NEW DELHI-110015

*Printed at*
D. B. Offset Works
B-218, Naraina Ind. Area
NEW DELHI

To my

MOTHER

whose intrinsic inspiration

helped me in completing this work

# FOREWORD

Dr. L.N. Bhagat's study of Supply Responses in Backward Agriculture (An Econometric Study), based as it is on a detailed econometric analysis of time series district-wise data on cereal production and allied variables, is an important contribution to the study of supply responses, in particular, and agriculture, in general, in a backward region like the Chotanagpur region with sizeable tribal population and forests. Besides providing district-wise and crop-wise supply response analysis in respect of substitution of crops on the one hand and inter-district comparisons on the other, the analysis is also carried out for the region as a whole on the basis of the panel data on the districts. Inter-temporal comparisons between the pre and post green revolution periods, probing the impact of the green revolution on supply responses, have also been made.

Apart from the usefulness of the study as a detailed, fairly disaggregate and fairly comprehensive econometric study of supply responses, four points stand out in Dr. Bhagat's work : (1) the insight which Dr. Bhagat, coming from the region, brings to bear from his first-hand knowledge of the region and its agriculture; (2) the wider connotation Dr. Bhagat gives to the concept of supply response; (3) the distinction that Dr. Bhagat makes between willingness of farmers to respond to price and growth signals and their ability to do so which depends on the infrastructure and financial and other support provided to them and is an important and perhaps more important than their willingness to respond; and (4) the heuristic explanation rather than a formally postulated explanation like the

Nerlovian one, offered as a more plausible explanation of the supply-response relationships Dr. Bhagat tests and the asymmetry in supply responses particularly in respect of subsistence crops that he notes and incorporates in his supply-response functions.

The interspersing of the main study based on official data with a few microstudies based on village surveys helps to supplement the main results and give the intensive micro-touch which the study would otherwise have lacked.

Department of Economics
Calcutta University
Calcutta-50
January 12, 1987

V. MUKERJI

# PREFACE

The study of supply responses in agriculture and their determinants on the one hand and constraints on the other becomes particularly important in the context of planning for economic development. Such studies in the case of backward economies with special problems and backwardness problem naturally need special attention. While studying the dynamics of supply responses in backward economies, it should be noted that the importance of such studies in backward regions lies not merely in testing the hypothesis as to whether the farmers are responsive to market stimuli, but also in raising and looking into the question whether adequate stimuli and support are provided to them for responding because farmers' response to price changes in a backward region is expected to be less marked than that in other advanced region due to a number of inherent difficulties. Various such other questions relating to the supply responses in backward agriculture deserve attention and careful investigation. Unfortunately, most of the Indian studies in this field have neglected the problems of backward agriculture and taken up the cases of advanced agriculture for their investigation.

The present study is an attempt to investigate the supply responses in agriculture and their differentials in a tribal concentrated Chotanagpur region - a most backward region of Bihar state in India. We have considered the concept of supply response in a broader sense, and various aspects of aggregate supply response and individual crop supply response have been examined. Acreage substitution behaviour between pairs of competing crops have been examined in detail and the overall cropping pattern response to price pattern also looked into. A simple analytical model has been developed for studying the supply responsiveness of the Chotanagpur farmers and the necessary statistical supports have been provided with.

The analysis covers six constituent districts of the Chotanagpur Plateau region and two different time periods. The first is the pre-green revolution era which stretches between the period from 1956-57 to 1965-66. The second refers to the post green revolution era covering the period from 1967-68 to 1976-77. The inter-district and inter-period analyses of the supply responses presented here are quite interesting and may be of some help to the planners at the lower level in identifying regional problems in traditional agriculture as well as in ushering in modern agriculture with its high yielding varieties and improved agricultural practices.

The present work forms a major part of the Ph.D. dissertation entitled "An Investigation of Supply Responses in Agriculture and their Differentials in a Backward Economy - Chotanagpur Region (1956-57 to 1976-77)" submitted to the Calcutta University in 1982. I am grately indebted to my supervisor Dr. (Mrs.) V. Mukerji, Professor of Economics, University of Calcutta for her exceptional interest in my work, valuable suggestions and consistent encouragement. My gratitude is due to several people with whom I held discussion on several occasions, especially Professor S.K. Bhattacharyya and Dr. A. Dasgupta of the Department of Economics, University of Calcutta, Professor N. Bhatacharya, Dr. R. Mukherjee and Dr. D. Kundoo of Indian Statistical Institute, Calcutta and Dr. G. Tribedy of Asutosh College, Calcutta. I am also thankful to Dr. Joseph Ryan (Jr.), a visiting American Professor for providing me some basic references with encouraging advices at the initial stage during 1977-78 when I have had an opportunity to work with him at St. Columba's College, Hazaribagh, Bihar.

The study was carried out during my stay at Calcutta during 1978-82 as a Teacher Fellow in the Department of Economics, University of Calcutta. The Fellowship was awarded by the U.G.C. New Delhi. The Vice-Chancellor, Ranchi University has kindly granted me academic leave for availing the fellowship. My thanks are due to them.

My thanks are also due to Dr. S.N. Sen, Joint Director, Directorate of Statistics and Evaluation, Government of Bihar, Patna and Sri S.C. Dutta, Deputy Director, Directorate of Agriculture, Government of Bihar, Patna for providing me with some relevant data. I am also thankful to the Head of the Department of Computer Science, Calcutta University, who has extended me the necessary facilities for the entire computational work.

I am grateful to Dr. Baldev Singh, Consultant of the Indian Council of Social Science Research who offered editorial advices and helped me in giving final shape of the book.

Finally, I offer my thanks to the Indian Council of Social Science Research, New Delhi for providing supporting grant for the publication of this book.

Department of Economics L.N. BHAGAT
Ranchi University,
Ranchi - 834008.

# CONTENTS

# LIST OF APPENDICES

# LIST OF MAPS

# LIST OF TABLES

# ONE

# INTRODUCTION

## 1.1 INTRODUCTION

In a country like India where agriculture constitutes the very basis of the national economy, agriculture deserves special attention while planning for development. A suitable policy is needed for proper utilisation of existing resources. The selection of proper policy, however, may be a matter of controversy because policy decisions relating to one sector of the economy may have conflicting implications for the other sectors. Sometimes even for a particular sector, a single uniform policy becomes insufficient due to the great variations in the socio-economic and agro-physical characteristics of the different regions in the country. This is why a greater emphasis is being given in recent years, on the study and adoption of regional problems and regional planning techniques. Regions with special problems and backwardness problems naturally need special attention. The present study, focussing its attention on some (the supply response) aspects in agriculture of a particularly backward region, namely, the Chotanagpur region, is an attempt in this direction. It may be of some help to the planners in identifying regional problems in traditional agriculture as well as in ushering in modern agriculture with its high yielding varieties and improved agricultural practices.

For proper planning and policy formulations in the agricultural sector, it is of paramount importance to understand the behaviour of farmers' supply responsiveness. Heady has noted that "The subject of farmers' responsiveness to economic variables has gained importance during the past few decades in both under-developed and developed economies. Developing economies need to understand the supply phenomena in order to implement policies for raising the output to a level which can provide for adequate human nutrition for the increasing population and also to promote general economic development. Even in developed countries, understanding of the supply phenomena is of crucial importance for controlling surplus, for raising farm incomes, and resource productivity" [55 p.3]. The study of supply responses in agriculture and their determinants on the one hand and constraints on the other becomes particularly important in the context of planning for economic development and the concomtant agricultural transformation and modernisation involved.

In this study we have examined the nature and extent of supply responses of major cereals in a backward region of Bihar in India for the period, 1956-57 to 1976-77. The period chosen can be divided into two broad equal sub-periods, the year 1966-67 being taken as the year from which HYV production was gradually ushered in along with other improved agricultural practices. The first period can be regarded as the pre-HYV period and the second as the post-HYV period. The period thus provides an opportunity to study supply responses in both the pre-HYV and post-HYV (hereafter referred to as HYV period) setups.

While studying the dynamics of supply responses in a backward economy, it should be noted that the importance of such studies in backward regions lies not merely in testing the hypothesis as to whether the farmers are responsive to market stimuli, but also in raising and looking into the question whether adequate stimuli and support are provided to them for responding because farmers' response to price changes in a backward region is expected to be less marked than that in other advanced regions due to a number of inherent difficulties like market imperfections, lack of infrastructural facilities and production of primarily subsistence crops, climatic uncertainties etc.

Economic theory suggests that prices are important determinants of economic behaviour and a rational farmer should sufficiently react to changes in the prices of output. But the peasant farmers of a backward economy practising traditional subsistence-agriculture may not respond to prices alone - some other factors may be more important for them - and they may not also respond to prices unless the changes in the prices are sufficiently large and persisting. Again the backwardness of their region and the socio-economic backwardness of the peasants there imply many infrastructural and informational shortages acting as constraints on their ability and motivation for responding to market-stimuli. The subsistence nature of their agriculture and the general non-monetised setup of their lifestyles may also dampen their need to be price-responsive. Due to their socio-economic attitudes and natural conservatism, the traditional farmers of a backward region poorly connected with the outside world may not require more income than a certain minimum that is necessary for their subsistence which in fact might lead to their responding to prices in a perverse way. Do these farmers actually behave in this way and does all this imply irrationality in their behaviour ? In a dynamic developing context as the present, how are the supply-responses changing ? Treating the innovative developments themselves as stimuli or incentives, are the farmers responding to them ? Is their price responsiveness increasing ? These are some of the questions which have been raised in this study and examined in the light of the data available at the district level for the farmers of Chotanagpur region.

## 1.2 DISCUSSION OF THE THEORY AND MODELS OF SUPPLY RESPONSES IN AGRICULTURE

### 1.2.1 Theory of Supply Response

Supply response in economic theory usually refers to the output production in response to their prices and the supply curves, that are anticipated, are generally derived from the assumption of profit maximisation. In agriculture, as in industry, farmers' decisions play an important role but the transformation-process involved depending as it does on a number of uncontrolled natural inputs along, of course, with the controlled physical inputs and human and animal labour, is very much more unpredictable than in industry. Every agricultural output is subject to biological processes that span a considerable period of time anything from one season to a calender year. Dharm Narain observes, "In agriculture, in particular, the biological nature of the production process makes for a considerable lag between production inputs and outputs. The lag between the time in decision in respect of utilisation of land, the most important input factor, and the fruition of that decision in output is still larger" [87 p.9]. The relationship between the farmers' decision and the resulting output is also not as definite and defined as in manufacturing industries due to the uncertainties caused by uncertain weather conditions. In fact farmers' decisions and hence direct or explicit responses can only be in terms of, space-wise and time-wise, land and input allocations and crop patterns though the output produced and or its value are the real targets and the resultant output appearing after a lapse of considerable period would depend on these but would not be fully determined by them.

It should be noted that profit maximising behaviour may not necessarily be present among the farmers. Farmers behave as producers but their decision as producers might also be affected by their decision as consumers of that crop, specially in the case of the subsistence crops. Actually the behavioural pattern determines the supply responsiveness of producers in a particular region and social and secular forces determine the level of commercial behaviour and hence the production-behaviour present among them. On the other hand according to Oury "Agriculture in purported to be the economic sector closest to perfect competition and therefore to classical economic theory, except for the usual assumption of complete mobility of factors of production and dependence upon weather" [99 p.13]. This may be true in so far as the atomic production set up in agriculture is concerned but in so far as producer-behaviour goes, the profit maximisation-principle of economic theory cannot be said to rule the roost completely.

In fact, the supply responses of an individual farmer involve the attitudes and decision-criteria of the farmer in chalking out his production-plans in the light of the anticipated returns from different crops. Thus the problem boils down to the question as to how the farmers allocate their land to different crops. Normally, that is for an economic farmer i.e., the crucial decisions relating to what crop and how much to grow on how much land with what inputs etc., are essentially based on the expected relative profitability of each crop under the different sets of inputs, ignoring for the moment the uncertainty element involved. Thus the composition and structure of the farm output keep changing over time as prices change. However, even in a subsistence economy with the present setup, the farmers generally sell a fraction of their output for purchasing other non-agricultural commodities for meeting their daily requirements like cloth, kerosene and salt etc. The prices of their output relative to the prices of other goods might influence the production-decisions of the farmers. Mellor [81] has observed that since a large fraction of production does not enter the exchange-part of the economy, changes in price relationships may appear to have no relevance in production-planning. However, particularly all farmers in low-income countries, produce at least a part of their produce for the market, and in turn get a portion of their consumption and production goods from the market. Such activities connect farmers with the market and make the relative price levels of goods sold and bought, of significance to the farmers and thereby provide potential for a price-effect on the volume of production. Mellor, however, goes on further to say "There is little or no evidence that farmers in lowincome countries operate according to a different set of guiding principles in making their managerial decisions than do farmers in high-incone countries. They do apply these principles in a different operating context. Consequently, they may often reach decisions quite different to those reached by farmers in high-income societies" [81].

Mellor appears to give a very broad content to his "operating context". One could as well have incorporated some of it in the production-behaviour-criteria and said that for these farmers profit-maximisation is not the only deciding criterion of production behaviour. One could replace "profit-maximising" behaviour by utility-maximising behaviour, utility now depending not only on material profit as understood in economic theory but also on other factors circumscribing the lives of farmers. But whichever way things are interpreted in the present setup even in backward regions, that prices would have some impact on supply response in agriculture cannot be denied.

It should also be noted that the role of the price mechanism in meeting the objective of allocative efficiency may be limited to a great extent by the physical, financial and institutional constraints prevailing in the backward economies. The choice of crop is frequently limited in view of the soil-climatic requirements and adequate irrigation facilities necessary for growing the crop. Traditional farmers of a backward economy might be practising crop-rotation which can be looked upon as systematic crop substitution and thus the decision of the farmer for planting the crop in alternate years may, to that extent, be quite independent of the prices. Under these circumstances, not only is the supply response to market prices restricted, but the market condition also loses much of its importance despite the prevalence of free entry into the grain trade. While all these arguments suggest that increased production is more dependent upon factors other than price, they do not mean, as Johnson [65] has pointed out, that prices do not matter very much.

However, a dynamic planning, developing context with changing technologies, changing infrastructural background and changing social motives, has two implications for supply response studies, one in respect of the extent of farmers' acceptance of the new technology and the other, in respect of their price or market responsiveness. Both responses could be regarded as two aspects of a generalised concept of supply-response. The changes in the infrastructural background and the social milieu may be able to usher in changes in basic attitudes resulting in changes in utility functions referred to earlier, leading to the gradual acceptance of the new technology on the one hand and improved market-responsiveness on the other, the two in turn feeding each other.

### 1.2.2 Models of Supply Response

In agricultural supply analysis, supply functions are often estimated directly by regression analysis of the time-series data, though cross-section data are sometimes used for production-function estimation. The supply curve can also be derived from empirically estimated produciton function under the assumption of known factor - prices and farmers' profit maximisation behaviour [26], but this method has been rarely followed. However, Wipf and Bawden [143] have made an attempt to evaluate descriptive and predictive reliability of the derived supply approach relating to United State agriculture. The authors derive firm-level supply functions from production functions for a variety of agricultural products and farm-types for various regions of the United States of America.

Supply elasticities and profit maximising output computed from these derived supply functions are compared with elasticities estimated directly by regression analysis and with actual output. The authors conclude that the empirical comparisons support the proposition that derived supply equations (derived from empirical production functions) are empirically less reliable.

Most of the previous studies in India deal with the problem of aggregate supply response of individual crops (with various degrees of aggregation over the group of individual decision makers) by taking planted area as a proxy for the planned output and compute acreage elasticity with respect to price as a measure of supply response. The basic model used for studying agricultural supply responses is that of the famous Nerlovian lagged adjustment model [92]. Most of the empirical research in Indian agriculture is based on either direct application or minor modification or further extension of the Nerlovian model. The final lagged relation which has been found empirically satisfactory in many cases can, however, be given alternative interpretation which could be better fitted to the backward economy cases (see section 1.6).

## 1.3 A BRIEF SURVEY OF PREVIOUS STUDIES ON AGRICULTURAL SUPPLY RESPONSE IN INDIA

A considerable body of literature has been developed on supply responses in agriculture in India, Africa, Thailand, Philippines, USA and other countries [6, 13, 33, 87, 92, 99, etc.] which includes the studies on individual crop as well as on groups of crops, on cash crops as well as on perennial crops including livestocks. While reviewing the previous literature we have restricted ourselves to the Indian studies relating to the case of annual crops specially cereals. Without going into a detailed critical examination of these studies we have briefly noted the procedure adopted, area covered and major findings obtained of these studies with a view to specify the major lacunae in the previous studies and develop alternative approaches.

In the Indian context Raj Krishna [71] was the first to use the Nerlovian model, in his study while examining the supply response of selected (food and cash) crops to price and non-price factors in the Punjab region during the pre-Independence period. The study was intended to put to the test the widely prevalent notion that peasants in poor countries did not respond, or respond very little or negatively to price movement. He observed that in the determination of acreage, price alone was the important factor identified in the case of maize and sugarcane; price was a more important factor than yield (in terms of elasticities) in the case of cotton (D); irrigation capacity was more important than price in the case of cotton (A), bajra and wheat irrigated; yield was more significant than price in the case of rice; and rainfall was most crucial in the case of unirrigated wheat, barley and gram. On the whole the supply response (as measured by acreage elasticity to price) of cereals was found to be less than that of cash crops. He has noted that no general presumption in favour of the irresponsiveness of crop output to prices in poor economies can be upheld. The responsiveness, however, varies as between different crops and regions.

Another important contribution in the field of supply response in India is the study by Dharm Narain [87]. Though the study is mainly based on the graphical method of analysis the author has attempted a rigorous exposition of almost all important aspects of the hypotheses concerning producers' behaviour in response to price changes in the context of peasant agriculture. The complex and intricate theoretical issues involved in assessing the role of basic economic principle of the profit-motive determining in the decision-making process of the Indian farmers have been thoroughly examined. He has concluded that the Indian farmers are significantly responsive to price incentives. The study reveals that the Indian cotton grower is not less responsive to commercial stimuli than his counterpart engaged in highly commercialised agriculture in the USA. It has been observed that in the case of foodgrain crops, rainfall assumes that status which price does in the case of cash crops. The study also demonstrates the role of price in foodgrain areas.

Maji, Jha and Venkataramanan [79] have studied the case of three major cereals - wheat, rice and maize in Punjab state for the period 1948 to 1965. In their study they have used a variant of Nerlovian area adjustment model, but they have taken note of Behrman's attempt [13] to include a measure of risk in the supply function. Standard deviations are taken as a crude measure of risk and taken as proxies for the variances of the subjective probability distributions. Both absolute and relative prices were tried in the response functions and consistently positive price-coefficients were observed for all the three cases. The response function of wheat, however, indicated that relative prices did not excercise a significant influence on acreage but absolute price proved to be significant. The inclusion of standard deviation of prices over the last three production periods as a measure of price-risk seemingly did not improve the explanatory power of the hypothesized relation, and the relevant coefficient though showing the expected negative sign was not statistically significant. The study indicated that large variations in prices had an adverse impact on acreage.

Jai Krishna and Rao [70] provide some more evidence in the field of supply response studies. They have investigated the possibility of alternative price formulations for studying supply responses, though the investigation is limited to wheat in Uttar Pradesh during the period 1950-51 to 1962-63. As much as twelve price formulations based on wholesale prices have been used to explain the changes in wheat acreage during the period under study. The results of the study indicate that the inference relating to acreage response changes drastically, depending on the estimate of the expected price used. In an another paper [107] they have used nine alternative price expectation models and six different response equations for analysing the dynamics of acreage allocation for wheat in Uttar Pradesh. Among the nine alternative price expectation models used in the study, the model based on three year average of pre-sowing price proved to be decidedly superior to the others. They also tested three non-Nerlovian models, identical to those but omitting lagged acreage (referred to as traditional regression model) and estimated supply elasticities and area adjustment coefficients resulting from regression based on the Nerlovian models. The results indicate that the traditional regression models give as satisfactory "if not superior" results as those obtained using the adjustment lag model of Nerlovian type. The Nerlovian type models were found to have an edge over the traditional models in so far as the proportion of variation in wheat acreage explained by the explanatory variables. The results of the study sufficiently indicate that it is possible to influence acreage allocation under wheat through the manipulation of relative prices of wheat and substitute crops.

As part of a major study of long term agricultural prospects in India, the National Council of Applied Economic Research [90] used a Nerlovian dynamic model to analyse the supply of cereals for the period from 1938 to 1957. The results on the whole indicate positive supply elasticities for rice, wheat, barley and jowar. In an another study NCAER [91] has also estimated the acreage responsiveness of four selected food crops viz. rice, wheat, jowar and gram to changes in prices for the period 1951-52 to 1964-65. The study reveals that the area of the four selected foodgrains responds little to the relative price variations. For the crops grown both for subsistence and for sale - wheat and rice in the surplus areas such as Andhra Pradesh, Madras and Punjab - the price elasticity of the area is in the medium range. The estimated short-run elasticities varies between 0.1 and 0.4.

Another important study in the field of supply response is the study by Acharya and Sengupta [1]. Considering the acreage substitution relation as an important aspect in the dynamics of supply the authors have examined the response of relative acreage to relative prices, in the case of two substitute crops viz. autumn rice and jute in the districts of Bihar and West Bengal. They observed insignificant price ratio coefficient in most cases and as an attempt to examine as to why the price sensitivity of acreage substitution is not as high as could be otherwise expected they have introduced risk element in their explanation. The coefficient of variation is taken as a rough measure of the risk and it is observed that the results are quite consistent with the theory of investment under risk. They argue that it is quite optimal for a subsistence farmer not to divert his acreage for jute (which involves much higher risk than the autumn rice), since this behaviour will be in keeping with the policy of playing safe under given risk conditions. Some other studies [63, 87, 118, 134 etc.] have also examined the relative acreage response to relative prices for pairs of individual substitute crops and established the importance of relative prices in acreage substitution between the crops. Mukerji & Mukerji [86] have, however, examined the substitution between superior and inferior foodgrains (group of crops) in India for the period from 1900 to 1958.

Madhavan [77] has studied the acreage response of four major cereal and four major commercial crops grown in Tamil Nadu for the period from 1947 to 1965. He observed that in almost all the cereal crop equations and even in some commercial crop equations, yield exceeds acreage elasticity with respect to price. Price coefficient-estimates show statistical significance in the supply of all cereals but rice. The author felt the rice regressions nevertheless showed cultivator responsiveness, since yield proved significant, and Madras government policies had been primarily aimed at increasing rice yields. The negative and significant price parameters indicated for cumbu (the best dry crop grown almost entirely for subsistence) also seemed logical to Madhavan. Though his results for both cereal and cash crops signified economic rationality in cultivators' allocative decisions, he cautioned that farmers are most likely to be restricted in their ability to respond to changes in the terms of trade by the practical realities of the region's overall situation one of which is that there is very little possibility of expanding the supply of land.

Nowshirvani [95] examined the planting behaviour of peasants in the districts of Bihar and the eastern half of Uttar Pradesh state, which are among the poorest areas in the subcontinent. He has studied the case of rice, wheat and barley among cereals with some other cash crops. Using modified Nerlovian model he has estimated the supply elasticities for two sub-periods (pre-war period 1909-1938 and post war period 1953-1962) for Uttar Pradesh but only for the post war period in the case of Bihar due to various data limitations. He observed very little supply response to price changes during pre-war period for rice crop in Uttar Pradesh. During the post-war period price parameters proved to show statistical significance in Bihar than in Uttar Pradesh. There was particularly no evidence of any significant supply response of wheat to price changes in either state. In both, the rainfall and trend variables show a somewhat greater influence on output than do wheat prices. In the case of barley, a somewhat greater likelihood of positive price responsiveness appeared in both Uttar Pradesh and Bihar. Summarising his findings, Nowshirvani pointed out the general insignificance of estimated price coefficients in the supply analysis of subsistence crops and the significance of the same estimate for the cash crops like groundnut and sugarcane. Commenting on the results obtained by Nowshirvani, Askari and Cummings have noted that "Nowshirvani's results pointing to the fact that the Nerlove model does not, in this case, give good positive price elasticity is one of the few major exceptions to such a finding. The Nowshirvani results could be justified on the grounds that the study was for a poor region in India and thus commercial behaviour could be absent [6 p.82].

Another study of cereals production in pre-Independence India has been recently reported by Parikh [101]. He has used the data collected by Dharm Narain [87]. Parikh observed insignificant price coefficient in most cases. A negative and significant price coefficient was observed for wheat in Madhya Pradesh/Berar region, but the coefficient of the deflated per acre proceeds (a measure of net proceeds per acre) was proved to be positive and significant, indicating the fact that the farmers responded positively to economic incentives, if not to the specific impulse of price. In estimates for Bombay/Sind, however, no significance was found for either price or net proceeds. Only Punjab cultivators showed positive price responsiveness. In all the cases the weather variable proved to have the most notable influence on wheat supply. Positive price response was found for Bihar/Orissa while negative and significant price, coefficients were observed for Madras. He, however, observed very high positive price response of the Indian farmers in the case of cash crops (for the same period). The generally weak performance of the various models tested for cereals perhaps indicates, more than anything else, the dominance of institutional factors generally unrelated to market agriculture as determinants of subsistence production.

Batra [11] has presented an important study of supply behaviour of peasant farmers in the Indian economy for the period 1952-53 to 1971-72. This study pertains to a dry "Subsistence" crop - bajra, an inferior quality cereal, in a drought-prone region of Gujrat. One of the major objectives of the study is to compare the acreage response of bajra under traditional technology (1952-53 to 1964-65) with its response under new technology (1966-67 to 1971-72). For this he has pooled the time series data of four selected districts and estimated the coefficients of selected independent variables using multiple regression analysis. The results indicate a crucial change in respect of the effect of price on acreage allocation under the crop in the two sub-periods. The price-coefficent in the acreage response equation has a positive sign under traditional technology but a negative sign under the new technology. The coefficients for the price variable are statistically significant in both the cases. The rationality behind the significantly negative price coefficient during the new technology period is examined, as the author argued, in terms of the substitution of HYV bajra for native varieties of bajra instead of increasing the total area under native bajra crop. The author observes that the bajra producers had no longer remained mere subsistence-producers during the new technology period.

As a part of the larger efforts at an identification of major social, economic and political factors which contribute to regional differences in market responsiveness, Cummings [27] has made a comprehensive district-wise study of supply responsiveness of Indian farmers. Using a modified Nerlovian model he has studied the acreage response to price and non-price factors of nine crops including three cereals - rice, wheat and barley. The evidence indicated a somewhat greater influence of prices on rice output than has been found in other studies of the subcontinent, though a majority of the estimated price parameters are negative or statistically insignificant. A considerably larger indication of positive price responsiveness is found for wheat than for rice in most cases. Positive price parameters dominate in the case of barley which is a minor crop and is commonly regarded as an inferior crop as compared to rice or wheat. Both rainfall and trend coefficient estimates are positive and significant in a majority of the districts. The overall results confirm in depth that farmers across the country are quite aware of the market and its potential reward, not only for cash crops, but generally as well for cereals.

Recently Cummings [28] has established the hypothesis that one reason why greater price responsiveness has been found for barley than for other cereals in several developing countries may be the generally inferior status of barley. He has examined the substitution between wheat and barley in 20 selected districts of Northwestern India. He observed that as the "Second choice" of many peasants higher barley prices may not only notably enlarge normally small plantings but also result in barley claiming somewhat better land.

In addition to the studies referred to above we have several others which analyse supply responsiveness of Indian farmers. A detailed summary of all such studies appears in some of the surveys of research works relating to the Indian economy like Shah [124] and

Bhagwati and Chakrabarty [21]. The majority view that emerges from these studies supports the hypothesis that the Indian farmers are responsive to market forces specially to changes in relative prices though this is less marked in the case of subsistence crops. A group of Indian economists meeting under the auspices of the Indian Society of Agricultural Statistics [32] had examined the general problems of price responsiveness. Basing their conclusions on the body of existing analytic evidence, the group concluded that farmers do in fact respond positively to prices, noting at the same time the existence of important and identifiable constraints that dampen the degree of responsiveness.

Recently Askari and Cummings [6] have made a world-wide survey of literature on agricultural supply responses (though restricting to the studies using Nerlovian model) and compared the results of nearly 200 studies of more than 75 agricultural products in some 40 countries. The study identified the price and non-price variables determining peasant behaviour in respect of production and marketing under differing conditions of geography, social structure and diverse crop production. From the comprehensive results presented in the Appendix-B one can easily arrive at the conclusions that (i) the farmers are responsive to market stimuli both in developed as well as in developing economies, (ii) there is a general tendency of cash crops to show greater responsiveness to price than food crops both in developed and in developing countries, (iii) the magnitude of response (measured in terms of acreage elasticity to prices) of cereals in developing countries is lower, in most cases, than that in developed countries, but there is ample evidence indicating that price responsiveness of farmers in developing countries is not less than that of farmers of advanced countries in the case of cash crops like cotton, sugar-cane, kenaf, tobacco and jute, (iv) among the cereals, mostly grown by peasant farmers in developing countries, minor crops show relatively higher price-response as compared to that of major ones.

On the basis of the review of literature on agricultural supply responses presented above we can identify some of the gaps in the existing literature. In the next section we have tried to point out some of them.

## 1.4 GAP IN THE EXISTING LITERATURE

The following are some of the major lacunae in the existing literature on agricultural supply responses :

(a) The major lacunae of supply response studies relating to Indian agriculture is, however, an absence of adequate number of studies of aggregate supply functions for farm products. The supply elasticity of aggregate agricultural output with respect to price is an important policy parameter in every country with a large agricultural sector and it takes an added interest where agriculture constitutes the very basis of the economy. Despite the recognition of its importance [30,72] very little is known of this crucial parameter in India. There are only a few studies in this line such as Herdt [57], Sawant [117] and Bapna [9]. However, there are numerous studies of this type in advanced countries specially USA. The most obvious reason for the apparent non-popularity of this type of analysis in India seems to be the conceptual, measurement and data problems.

(b) In supply response studies, usually planted area is considered as a proxy for planned output and computed acreage elasticity with respect to price as a measure of supply response. In fact the acreage response to price is, but only one of the facets of the total output response. In modern agriculture with increasing use of new agricultural practices, the farmers' response to price changes may also be reflected in terms of increased area under the crop. Increase in area under the crop becomes, sometimes, very difficult due to agro-climatic requirements needed for growing the crop and limited availability of land. Farmers' response to price changes may also be reflected in terms of changes in cropping pattern and cropping intensity. The proper utilisation of land, specially the utilisation of fallow land (wherever exists) may also takes place as a result of a rise in prices. Adoption of new agricultural practices may, itself, reveal farmers' responsiveness. In a generalised sense in supply response one might also include behavioural supply response. Here by behavioural supply response we mean changes in the attitudes and decision-criteria underlying production behaviour of the farmers both in normal time and in times of natural calamities like severe drought or severe floods. Any efforts of the farmers for compensating the loss of output in a season/year caused by monsoon failure, in subsequent season/year, provided normal rainfall occurs, might indicate one such behavioural response in calamitous circumstances.

(c) As Askari and Cummings have noted, in their recent survey of literature on agricultural supply response, "The results give price elasticities but nothing or nearly nothing about the conditions that determine the magnitude of these elasticities. Many authors, for example, in reporting on two or more crops and/or regions, have speculated about the reasons for any differences they found, to be sure, but none have attempted any systematic verification of such explanations ..... Thus they have included little more substantive evidence regarding an explanation of differences, across crops or regions, of the calculated elasticities than have the briefer works" [6 p.342].

(d) Most of the studies are macro in nature and refer to the analysis of commercial crops of relatively advanced region or state. For instance most of the Indian studies refer to the Punjab state, the most agriculturally advanced state in India, whereas in Bihar or Orissa, the most backward states, such studies are relatively few. There is a need for micro studies at lower levels. All India level studies or even at the state level analyses do not throw light on the problems of some isolated cut-off backward regions which might really exist within the state. The characteristics of decision-making units and production conditions in such isolated economies might be different from those in advanced economies and the explanation of their behaviour might also require some alternative framework of analysis.

## 1.5 OBJECTIVES AND PLAN OF WORK

The present study is mainly based on secondary sources of data at district level. The main objective of this study is to investigate the nature and extent of supply response of important cereals (individually as well as in the aggregate) in the Chotanagpur region, a socially and economically backward region of a backward state (Bihar) in India, and probe into its (supply response) inter-district and intra and inter period variations over the period and their causes.

In this study we have considered the concept of supply response in a broader sense and various aspects of aggregate supply response and individual crop supply response for important crops have been examined in detail subject to the availability of data. Acreage substitution behaviour between pairs of competing crops have been examined in detail and the overall cropping pattern response to price pattern also looked into.

The period under investigation stretches from 1956-57 to 1976-77. Considering the important requirements like consistency, comparability and reliability of data in mind, we have decided to begin our study from 1956-57. The upper limit is restricted by the most recent availability of data (Detailed analysis of the sources of data and criterion for selecting the period is given in Appendix-A at the end of the book). The available data have been split up into pre-HYV (1956-57 to 1965-66) and HYV (1967-68 to 1976-77) period. The year 1966-67 was the year of the introduction of HYV technology in Bihar. The particular year 1966-67 has been left out for the purpose of time series analysis as there was severe drought in that year and on this ground it has been treated as an abnormal year.

The study is confined to the six districts of Chotanagpur region viz. Santhal Pargana, Hazaribagh, Palamau, Ranchi, Singhbhum and Dhanbad. It should be noted that Hazaribagh district ( the only district in the region affected by the 1972 division of districts[1]) in the present study refers to the pre-1972 Hazaribagh. The 1972 division gave rise to a new district "Giridih" within the same geographical area. Necessary adjustments have been made to maintain the consistency of data. Various other adjustments of data have also been made (Detailed analysis of the necessity and methods of adjustment of data is given in Appendix-B at the end of the book).

Chotanagpur plateau region is one of the three natural divisions of Bihar with a number of distinctive characteristics distinguishing it from the rest of Bihar. It is marked by hilly, uneven territory and a predominance of forests and tribals. Its agriculture is backward and

---

1. The number of districts in Bihar was only 17 before 1972. Since then many new districts have been created for better administrative reasons and at present the number of districts is 37.

based primarily on subsistence crops and dry-farming. Inadequate and uncertain rainfall, lack of irrigation facilities (or at least insufficient irrigation facilities) fertilisers and the hilly uneven terrain and the social backwardness of the tribal population have naturally led to low yield rates and concomitant poverty and along with transport difficulties, market imperfections and the existence of various institutional and cultural constraints acting as bottlenecks have led the region and its agriculture in to stagnation and what may be called a backwardness-trap. There is, however, as is well known an abundance of minerals and forest wealth. Mining and coal-based industries have also been built up but these are confined to some urban pockets. Their prosperity has not adequately trickled down to the rest of the area. The forest wealth has also not been properly utilised and developed. It is only recently that serious efforts at breaking this backwardness-trap and bringing about overall development in the region are being made among other things in the form of increased irrigation facilities, availability of new agricultural practices and utilisation of forest resources.

As mentioned earlier two additional points may be noted in the context of the Chotanagpur region. Though the Chotanagpur region has tribal concentration, the socio-cultural attitudes of the tribals of Chotanagpur significantly differ from other Indian tribals in many respects. Chotanagpur tribals are less rigid in their attitudes and have attained a greater degree of social change [116]. The other points to be noted are that though Chotanagpur region as a whole has distinctive characteristics as compared to the rest of Bihar, the districts of Chotanagpur region differ among themselves in respects of many characteristics such as tribal concentration, irrigation facilities, cropping pattern, their mineral and forest endowments, urbanisation, industries etc..

The Chotanagpur farmers mostly grow cereals and that too for their own consumption; however, a fraction of that is sold for purchasing other goods of daily necessity. Rice is the most important crop grown by almost all farmers, with coverage varying from 31 per cent in Palamau to 80 per cent in Singhbhum. Due to undulating topography, existence of hills and forests the percentage of net area sown varies between districts from about 19 per cent in the Hazaribagh district to 40 per cent in the Santhal Pargana district.

The cropping pattern is biased towards cereals with rice, maize and ragi grown during kharif season and a very small proportion of cultivable area (only about 14 per cent) is used for growing rabi crops mainly wheat, barley and gram during the rabi season. An interesting and important feature of this region is that all holdings are owned and self operated in all districts with the exception of Hazaribagh as per the 1971 agricultural census. It is pertinent to note here that the agro-economic conditions for the major cereals grown in the region differ between the crops.

The traditional method of cultivation which must have emerged out of their experiences over the ages is still predominating among the farmers of Chotanagpur. Farmers mostly use non-marketed inputs like their own traditional wooden plough drawn by a pair of bullocks, own seed, own fertiliser (i.e. organic manures) and a few other simple tools. After the introduction of the HYV technology in Bihar in 1966-67, the farmers of Chotanagpur have also been participating and sharing the benefits of green revolution to some extent. However, late adoption of new agricultural practices by the Chotanagpur farmers has been reported which could be due to various social, economic, institutional and informational constraints [16]. The average adoption index of new agricultural practices for Chotanagpur region (see item 28 in Table 2.1) is also very low as compared to the corresponding average adoption index of plain regions of Bihar. At the village level, however, there is sufficient evidence to indicate that the farmers of Chotanagpur including the tribal farmers are gradually using these practices for growing wheat [16].

We have also worked out the direction and magnitude of changes if any in the different agricultural variables relevant for the supply response study and the inter-district and inter-period differences noted. The latter, apart from providing a factual background and profiles of the districts relevant for the study, help in formulating the hypotheses and understanding the nature and determinants of the supply responses.

Thus owing to the observed differences in the socio-economic and agro-industrial characteristics of the districts in the region and changes therein over the period, our main thrust in this study is to investigate

(a) Differences in the price responsiveness of supply of different crops and of the aggregate of all the crops within and between the districts in the two sub-periods (i.e. pre-HYV and HYV period) and factors underlying them.

(b) Differences in the price responsiveness of substitution among the substitute crops within and between the districts in the two sub-periods and factors underlying them.

(c) Impact of new technology on supply responses and

(d) Relative importance of price and non-price factors in determining the supply responses in the region.

## 1.6 THE MODEL USED FOR STUDYING SUPPLY RESPONSES IN THE CHOTANAGPUR REGION AND THE HYPOTHESES MADE

Owing to the peculiar characteristics of the decision-making units and also the peculiarities of the production-conditions under which the Chotanagpur farmers take decisions, we hypothesize that the farmers may not be consciously formulating a precise quantitative decision-pattern but broadly similar surrogate considerations may be at work behind their decisions.[2] Their lack of formal ability, information and control over the situation on the one hand and trust in the time-tested age old agricultural patterns evolved and planned as a stable system on the other may result in their following some simple easily understood working rules characterised by conservatism, traditionality, risk aversion, subdued response to changes and

---

2. Such considerations, of course, should not preclude explanatory or normative theoretical constructs broadly consistent with or suitable for the actual psychological or social processes involved. Existing economic theory abounds with such theoretical constructs. Such quantitative models have their own value in that they reflect theoreticians' attempts to understand the behaviour-pattern and give it a formal formate. In this quest the theoreticians may not only be successful in identifying appropriate patterns but also in using them or developing new patterns for forecasting and normative purposes.

responsiveness to relatively persistent and relatively large changes.

Therefore we have used a simple model for studying supply response in Chotanagpur region which is based on the hypothesis of " a simple working rule " i.e. the amount of land given to an individual crop (s) in one period depends on the area allotted to the crop (s) in the previous period along with some adjustments in response to changes in prices and other factors. The hypothesis of the above simple working rule leads to the heuristic formulation that acreage planted under the crop in the current period is a function of area planted in the previous period and the price of the crop in the previous period. An analytical model is developed which enables us to identify suitable shifter variables (see section 4.6) for acreage response studies. Our models takes into account the technical and environmental factors affecting supply responses.

It should be noted, however, that our simple model is apparently similar to the Nerlovian partial adjustment model in its final derived form, the only difference lying in the generating process of the lagged dependent variable and its coefficients. In the Nerlovian framework it is the outcome of the adjustment mechanism assumed in the model whereas we have taken it as a traditional variable, which infact reveals some interesting traditional characteristics of the farmers in the region.

Some of the explicit concomitant hypotheses which have been examined in the course of the analyses are the following :

(1) Farmers' responsiveness to price changes is limited which may be because of the various agricultural constraints operating on them.

(2) Farmers' response is not symmetrical in the sense that it is different when price increases and when price falls. We call this as subsistence reservation hypothesis.

(3) Non-price factors are more important than the price factor in the case of the subsistence crops.

(4) Peasant farmers are risk-averters rather than risk-takers.

(5) Response pattern remains almost the same in the two sub-periods (i.e. in the pre-HYV period and in the HYV period) except in the case of wheat for which the use of HYV seeds in the HYV period has been quite marked.

## 1.7 CHAPTER SCHEME

**The Chapter scheme of the book is as follows :**

Chapter 1, as is already evident, is introductory in nature. Besides a brief survey of the previous studies relating to the analysis of supply responses in Indian agriculture, it introduces some of the important aspects of supply responses which may be necessary for studying aggregate, individual crop and substitution supply response in agriculture. Some of the major lacunae in the existing literature are briefly discussed and the main stand of the present study and the hypotheses to be examined in the course of the analysis presented.

Chapter 2 is mainly concerned with the background profiles of the Chotanagpur region which is selected for the present study. It examines the extent of backwardness of the Chotanagpur region as compared to the other regions of Bihar and socio-economic and agro-industrial characteristics of the districts in the region are discussed in detail. It also provides some agro-economic characteristics of the major crops grown in the Chotanagpur region.

In Chapter 3 we have examined the behaviour of agricultural output, input and prices in the districts of Chotanagpur region in the two sub-periods, which provides us the basis for determining the tentative hypotheses for supply response studies. The behaviour of agricultural output, input and prices are examined in terms of the semi-log-linear trend equation fitted for each districts and for the two sub-periods for the relevant items. We have also estimated the extent of fluctuations around the trend line specially of area, productivity and output, in terms of $1\text{-}R^2$ values where $R^2$ is the multiple correlation coefficient, in the two sub-periods for important crops. $1\text{-}R^2$ values which have been taken as a measure of the fluctuation around the trend line could be taken to reflect the uncertainty and risk involved in cultivation.

The subjects of Chapter 4 are the models and estimation techniques which are used to obtain estimates of the supply responses of Chotanagpur farmers. It introduces some special characteristics of the traditional agricultural system and their rationality in the Chotanagpur region and the resultant characteristics of the agricultural supply responses for the region. The simple working rule hypothesis has been offered and a simple analytical model is developed for studying the behaviour of supply responses of Chotanagpur farmers. The basis for the use of planted area as the dependent variable and other independent variables considered for the purpose is discussed and alternative specifications of the multivariate regression equation are presented and analysed.

Chapter 5 is concerned with the acreage response of Chotanagpur farmers with respect to price and non-price changes. It has been examined at two levels, firstly at the individual crop-level and secondly at an aggregate level i.e. all cereals crops taken together. At an individual crop-level we have examined the case of major cereals viz. winter rice, autumn rice, maize and ragi grown during kharif season and wheat and barley grown during rabi season. The results are presented and analysed within the framework of various a priori hypotheses. Analysis is carried out for the two sub-periods and also for the combined period in each district and in the Chotanagpur region as a whole after pooling the districts data. Non-symmetric acreage response to price changes has been hypothesized and examined in the case of two major cereals winter rice, the major kharif cereal and wheat, the major rabi cereal, in the districts of Chotanagpur region for the combined period. The Chow statistic is used for testing the overall homogeneity of regressions in the two sub-periods. The differences in the price-responsiveness over time and space (i.e. district) is examined and the possible causes behind these differences are analysed.

In Chapters 6 and 7 we have examined the behaviour of the cropping pattern response to price pattern in the districts of Chotanagpur region in the two sub-periods as well as for the combined period data. Chapter 6 is devoted to the discussion of crop substitution behaviour of pairs of competing crops in two important seasons (kharif season and rabi season) separately for the two sub-periods and also for the combined period. Standard statistical tests are applied for testing the overall homogeneity of regressions in the two sub-periods. In Chapter 7, an attempt is made to estimate the magnitude of changes in the overall cropping pattern and factors underlying these changes in terms of changes in the overall price pattern and other relevant factors over the period in the districts of Chotanagpur region.

In Chapter 8 a summerized version of the overall findings and conclusions are presented. Some limitations of the present study and suggestions for further research work have also been noted. At the end of the book, we have given as an Appendix, sources of and other notes about the data (such as selection of period and adjustment of data) used in the study.

TWO

# CHOTANAGPUR REGION

## 2.1 INTRODUCTION

The Chotanagpur plateau region which is an economically and socially backward region of a backward state (Bihar) is selected for the present study. Bihar has been divided into three natural regions [3] considering soil characteristics : (i) North Bihar Plain (NBP) region located in the north of the river Ganges with newer alluvium as the major soil category, (ii) South Bihar Plain (SBP) region located in the south of the Ganges with older alluvium as the major soil category, and (iii) Chotanagpur Plateau (CNP) region located in extreme south and south-east of Bihar with sedentary type soil. The word plateau is a technical expression for the area whose lowest level is at a considerable height above the sea. The seventeen undivided districts of Bihar - this refers to the pre-December 1972 period; at present the number has gone to 37 due to the formation of new districts - have been divided into three natural groups as under

NBP : Comprising the districts of Saran, Champaran, Muzaffarpur, Darbhanga, Saharsa and Purnea.

SBP : Comprising the districts of Patna, Gaya, Shahabad, Monghyr and Bhagalpur.

CNP : Comprising the districts of Santhal Pargana, Hazaribagh, Palamau, Ranchi, Singhbhum and Dhanbad.

Thus Chotanagpur is one of the three natural regions of Bihar consisting of 6 districts (see Map 1). It should be noted that Hazaribagh district (the only district in the Chotanagpur region affected by subsequent division) in the present study refers to the pre-1972 undivided Hazaribagh district - its division in 1972 has given rise to a new district "Giridih" within the same geographical area. Necessary adjustments have been made to maintain the consistency of data over the period studied.

## 2.2 HISTORICAL SKETCH OF THE CHOTANAGPUR REGION AND SELECTION CRITERION

Chotanagpur is a modern name given to the plateau region of Bihar which is bounded in the north by South Bihar Plain, on the south by the state of Orissa, on the east by the state of West Bengal and on the west by the state of Madhya Pradesh. We hardly find any reference to this region in our ancient literature. According to traditional accounts there were two important kingdoms in Bihar at the time of Bharat war, the kingdom of Magadha embracing the districts of Patna and Gaya in South Bihar and the kingdom of Videha, comprising probably of the old kingdom of Visala or Vaisali in North Bihar [48]. The Jain literature, however, throws some light in this direction but is limited to a few places only such as Parasnath hill, where a few ascetics went to meditate in peace and solitude and Pasubhumi, conquered by Bhima [100]. Parasnath hill is located in the Hazaribagh (now it lies in Giridih) district. Pasubhumi has not been identified but one may, however, locate it somewhere in modern Chotanagpur because there are districts like Singhbhum, Manbhum (former Dhanbad) and the like besides the similarity of the flora and founa characteristics of this region with that described in literature in the context of Pasubhumi.

Because of the tribal concentration in this region it is also called Tribal Bihar. "The word Jharkhand has been used for this region in later sanskrit literature but we cannot actually ascertain when it was first named thus. The Muslims call this region Kokrah. There were no kingdom of significance in this area. The petty chiefs who occupied small lands belonged generally to the aboriginal tribes. Local traditions declare that they were Nagavansi chiefs, which is sufficient to indicate their descent from wild tribes" [100p. 111].

The region is essentially a series of plateaux which descend from the main plateau of Ranchi (Netarhat) towards the, south and east. Except the main plateau of Ranchi with 2000′ to 3000′ elavation most of the area is above 500′ above the sea level. Only the eastern part of Dhanbad and Santhal Pargana districts has less than 600′ height. The area is full of hills, valleys and sloping lands, drained by several large rivers, such as Damodar, Baraker, Subarnrekha, North Koel, South Koel, Sankh, Baitarni and Mahanadi. The area is subjected to heavy erosion due to surface run-off which has resulted in the loss of top soil and depletion of fertility, formation of rills, gullies etc. The region is still largely covered by forest and is thinly peopled. The whole area belongs to the same geographical formation. The selection of the region for the present study is based on the following grounds:

(1) Homogeneity of land is almost assured [3]. The red and yellow light soil mainly composed of sandy clay covers the whole area with some laterite soil in south-west part of Ranchi and south-east part of Singhbhum and black soil in eastern part of Santhal Pargana (see Map - 2).

(2) It is a hilly and uneven region with relatively backward economy with concentrated tribal population. The backwardness of this isolated region is also borne out by the fact that a separate Chotanagpur-Santhal Pargana Autonomous Development Authority was set up at Ranchi as per Bihar Act 8, dated November 13, 1971 ; this has now been replaced by three sub-authorities : North Chotanagpur, South Chotanagpur and Santhal Pargana Autonomous Development Authority with headquarters at Hazaribagh, Ranchi and Dumka respectively as per Chotanagpur-Santhal Pargana Autonomous Development Authority (Second Amendment) Ordinance 1978 vide Bihar Ordinance No. 59, dated June 5, 1978. Besides this, the office of the Regional Development Commissioner has been established by the Planning Department, Government of Bihar at Ranchi in November 1975 to formulate and implement the regional plan for Chotanagpur region with the sub-plan provision in Bihar state plan.

(3) The districts differ from each other in many respects such as, area under forest, net area sown, irrigated area, industrial develop-

ment, tribal concentration, literacy rate, average farm-size, land concentration, road length and even in their relative backwardness and many others.

(4) A relatively small geographical area of the region was another consideration for this would make it possible to formulate the actual alternatives faced by the farmers and identify the appropriate shifter variables relevant for the supply response functions.

## 2.3 SOCIO-ECONOMIC BACKWARDNESS OF THE CHOTANAGPUR REGION AS COMPARED TO OTHER REGIONS OF BIHAR

A detailed analysis of the relative backwardness of Bihar among the states in India has already been made by Prasad [103]. Actually Bihar itself exhibits many inter-regional differences. Regional characteristics indicate the fact that the plateau region of Chotanagpur is socially and economically backward as compared to the other plain regions of Bihar. Some socio-economic characteristics of these regions are examined below in Table 2.1. Apart from the 1971 census data and data from a few other specific studies used for the purpose, the data in Table 2.1 refer to the year 1975-76, latest year for which published data were available. These data (in Table 2.1 showing regional disparities in Bihar) reveal the following features of the Chotanagpur region :

(1) The Chotanagpur plateau region, lying between 22°N and 25.3°N latitude and between 83.3°E and 88°E longitude with tropical humid type climate, covers about 46 per cent of the total area and about 25 per cent of the total population in the state, implying that the region is thinly populated as compared to other regions of Bihar. Rural-urban ratio is 84:16 in Chotanagpur region as compared to 95:5, 88:12, and 90:10 for North Bihar Plain, South Bihar Plain and Bihar state respectively. So far as general literacy rate is concerned, Chotanagpur is placed next to South Bihar Plain but the rural literacy rate is the lowest (15.13%) in Chotanagpur among the regions in Bihar as per 1971 census.

(2) All the three regions are basically agricultural as indicated by the very high proportion of working population engaged in the

primary sector (see item 5 in Table 2.1). The percentage of workers engaged in the secondary sector is highest in Chotanagpur (12.63%) because of industrial concentration in Dhanbad and Singhbhum districts. Percentage of workers engaged in tertiary sector is highest in South Bihar Plain (12.38%) followed by Chotanagpur (11.75%) and North Bihar Plain (7.72%) region.

TABLE 2.1 : **Showing inter-regional disparities in Bihar (Some selected indicators)**

| Characteristics /Region/State Plain Region | North Bihar Plain Region | South Bihar Plateau region | Chota nagpur | Bihar State | |
|---|---|---|---|---|---|
| 1 | 2 | 3 | 4 | 5 | |
| A. General Characteristics | | | | | |
| 1. Total Geographical area by village paper in thousand hectares [42] | 4910 (28.33) | 4450 (25.68) | 7970 (45.90) | 17330 (100.00) | (i) & (ii) |
| 2. Total population in lakhs as per 1971 Census [42] | 241.89 (42.92) | 179.37 (31.83) | 142.27 (25.25) | 563.53 (100.00) | |
| 3. Rural-urban ratio as per 1971 Census [42] | 95:5 | 88:12 | 84:16 | 90:10 | |
| 4. Literacy rate in percentage as per 1971 Census [42] | | | | | |
| i) Rural literacy rate | 15.69 | 20.85 | 15.13 | 17.17 | |
| ii) Urban literacy rate | 32.48 | 52.62 | 45.35 | 44.92 | |
| iii) General literacy rate | 16.91 | 23.67 | 20.39 | 19.94 | |
| 5. Percentage distribution of working population as per 1971 Census [42] | | | | | |
| i) Primary sector | 89.01 | 81.66 | 75.62 | 83.22 | (iii) |
| ii) Secondary sector | 3.27 | 5.96 | 12.63 | 6.55 | |
| iii) Tertiary sector | 7.72 | 12.38 | 11.75 | 10.23 | |
| 6. Percentage of socially backward people as per 1971 Census [42] | | | | | |
| i) Scheduled Tribe (ST) | 0.79 | 0.96 | 32.12 | 8.75 | |
| ii) Scheduled Caste (SC) | 13.91 | 17.68 | 9.93 | 14.11 | |

*Contd.* Table 2.1

| 1 | 2 | 3 | 4 | 5 | |
|---|---|---|---|---|---|
| iii) Total of ST & SC | 14.70 | 18.65 | 42.05 | 22.86 | |
| B. Land Utilisation | | | | | |
| 7. Area under forest in thousand hectares during 1975-76 [42] | 83 (1.69) | 518 (11.64) | 2220 (27.85) | 2821 (16.28) | |
| 8. Land not available for cultivation in thousand hectares during 1975-76 [42] | 929 (18.92) | 1264 (28.40) | 3321 (41.17) | 5514 (31.82) | (iv) |
| 9. Culturable waste land (CWL) in thousand hectares during 1975-76 [42] | 683 (13.91) | 639 (14.36) | 2027 (25.43) | 3349 (19.32) | (v) |
| 10. Net area sown (NAS) in thousand hectares during 1975-76 [42] | 3297 (67.15) | 2547 (57.24) | 2623 (32.91) | 8467 (48.86) | |
| 11. Gross area sown (GAS) in thousand hectares during 1975-76 [42] | 4748 | 3597 | 2924 | 11287 | |
| 12. Percentage utilisation of the cultivable land during 1975-76 [42] | 82.84 | 79.94 | 56.41 | 71.66 | (vi) |
| 13. Cropping intensity (CI) during 1975-76 | 1.44 | 1.41 | 1.12 | 1.33 | |
| C. Land Concentration | | | | | |
| 14. Concentration ratio during 1976-77 [113] | 0.530 | 0.519 | 0.604 | 0.574 | (vii) |
| 15. Average size of operational holdings (in hectares) during 1976-77 [113] | 0.84 | 1.01 | 1.91 | 1.11 | |
| D. Irrigation Facilities | | | | | |
| 16. Percentage of total area sown irrigated during 1975-76 [42] | 23.46 | 54.77 | 9.35 | 29.76 | |

*Contd.* Table 2.1

| 1 | 2 | 3 | 4 | 5 | |
|---|---|---|---|---|---|
| 17. Source-wise distribution of total area irrigated during 1975-76 in thousand hectares [42] | | | | | |
| (i) Canals (28.90) | 322 (41.47) | 817 (14.91) | 41 (35.13) | 1180 | (viii) |
| (ii) Tanks (3.95) | 44 (2.03) | 40 (18.91) | 52 (4.05) | 136 | |
| (iii) Tube Wells (47.31) | 527 (26.14) | 515 (4.00) | 11 (31.35) | 1053 | |
| (iv) Other Wells (7.72) | 86 (5.03) | 99 (31.64) | 87 (8.10) | 272 | |
| (v) Other sources (12.12) | 135 (25.33) | 499 (30.54) | 84 (21.37) | 718 | |
| Total (gross) area irrigated from all sources | 1114 (100.00) | 1970 (100.00) | 275 (100.00) | 3359 (100.00) | |
| E. <u>Some other Allied Facilities</u> | | | | | |
| 18. Availability of agricultural credit (in Rs.) through various agencies (viz. Government, Co-operatives & Commercial banks) per rural family during 1976-77 [37, 111] | 45.24 | 65.79 | 23.43 | 46.27 | |
| 19. Length of road (total) in Kilometres per thousand square kilometres of area as on March 31, 1975 [42] | 796.14 | 476.84 | 390.79 | 540.80 | |
| 20. Number of composite demonstrations for paddy and wheat per block during 1976-77 [42] | 19.40 | 18.55 | 11.46 | 16.54 | (ix) |

*Contd.* Table 2.1

| 1 | 2 | 3 | 4 | 5 | |
|---|---|---|---|---|---|
| 21. Number of regulated markets per block as on October 29, 1975 [37, 41] | 0.218 | 0.185 | 0.088 | 0.162 | |
| F. Seasonal Cropping Pattern : | | | | | |
| 22. Percentage distribution of gross area sown (GAS) between different seasons during 1975-76 [42] | | | | | (x) |
| (a) Kharif Season (June to December) | 58.49 | 50.13 | 86.13 | 63.03 | |
| 1 | 2 | 3 | 4 | 5 | |
| (i) Bhadai | 22.58 | 9.29 | 23.66 | 18.62 | |
| (ii) Aghani | 35.91 | 40.84 | 62.47 | 44.41 | |
| (b) Rabi Season (November to May) including garma crops and orchards | 41.51 | 49.87 | 13.87 | 36.97 | |
| G. Use of New Inputs in Agriculture : | | | | | |
| 23. Total value of machines used for agricultural purposes (in Rs.) per hectare of net sown area during 1976-77 [37, 39] | 169.36 | 186.66 | 30.93 | 134.58 | (xi) |
| 24. Total consumption of chemical fertilisers (NPK) per hectare of net area sown during 1976-77 (in Kg. nutrient) [39, 137] | 18.75 | 31.99 | 4.27 | 18.65 | |
| 25. Percentage of area under HYV seeds to the total area under [37,42] | | | | | |
| i) Paddy | 12.76 | 19.73 | 7.05 | 12.71 | (xii) |
| ii) Wheat | 77.63 | 55.74 | 60.80 | 66.86 | |
| iii) Maize | 75.82 | 37.72 | 13.05 | 50.46 | |

*Contd.* Table 2.1

| 1 | 2 | 3 | 4 | 5 | |
|---|---|---|---|---|---|
| iv) Total of Paddy, Wheat & Maize | 37.29 | 32.82 | 10.10 | 28.98 | |
| H. Miscellaneous : | | | | | |
| 26. Agricultural productivity (in Rs. per hectare at 1961 prices) during 1976-77 [39,42] | 387.11 | 577.91 | 321.89 | 435.52 | (xiii) |
| 27. Composite index of infrastructure for the year 1976-77 (regional averages) [37,42] | 92.54 | 143.88 | 50.51 | 100.00 | (xiv) |
| 1 | 2 | 3 | 4 | 5 | |
| 28. Adoption (of new agricultural practices) index for the year 1976-77 (regional averages) [37,42,137] | 110.43 | 130.02 | 36.20 | 100.00 | (xv) |
| 29. Gross value of agricultural output (in Rs. at 1961 prices) per capita of rural population during 1976-77 [39, 41, 42] | 79.82 | 103.12 | 56.38 | 81.55 | |
| 30. Per capita annual income from all sources in rural areas during 1976-77 (regional averages) [44] | 779.52 | 774.81 | 732.16 | 766.50 | (xvi) |
| 31. Overall economic development (simple) scores for the year 1974-75 (regional averages) [43] | -2.87 | +24.60 | +5.04 | 0.0 | (xvii) |

Supplementary notes :

(a) Figures in square brackets in the first column are the numbers allotted in the bibliography and indicate the source (s) of information for the respective item.

*Contd.* Table 2.1

(b) The figures in round brackets in each of the columns 2, 3, 4, 5 (wherever given) stand for the corresponding percentages.

(c) Figurres in round brackets at the end of the fifth column are the numbers allotted to the remarks related to the corresponding item and data in that row and given below.

(i) Regional total and percentages are computed.
(ii) Total may not tally due to rounding off figures.
(iii) Items included in broad categories are -

| | |
|---|---|
| Primary sector : | Cultivators, agricultural labourers, livestock, forestary, fisheries and other allied activities. |
| Secondary sector: | Mining & quarring, household and other than houschold industries. |
| Tertiary sector : | Construction, trade & commerce, transport & communication and other services. |

(iv) Land not available for cultivation includes forest land, barren land and land put to non-agricultural uses.

(v) Culturable waste land includes cultivable waste other than fallow land, fallow land (current as well as other fallow lands), permanent pasture & other grazing land, and land under miscellaneous trees and groves.

(vi) Percentage utilisation of cultivable land $= \frac{NAS}{NAS + CWL} \times 100.$

(vii) Concentration ratios are computed using the proportion of number and area of operational holdings by different size-groups (broad categories) of operational holdings.

(viii) Total (regional & S ate) figures are the horizontal sum of the districts' (rounded off) figures given in the source. Other sources of irrigation includes pyn, aahars, etc.

(ix) Composite demonstrations are conducted by the officials of the Department of Agriculture, Government of Bihar on farmers plot showing the combined effect of using the new agricultural practices mainly chemical fertilisers, HYV seeds and insecticides.

(x) Winter rice is a major aghani crop whereas autumn rice, maize and ragi are called bhadai crops. Wheat, barley and gram are major crops grown during rabi season.

(xi) Machines included in this study are tractors, power tillers, pumping sets (both diesel and electric), sprayers and dusters and the value of machines is evaluated as per average market prices.

*Contd.* Table 2.1

(xii) Percentage refers to the 3 years' average percentage and computed from the actual values of area under HYV seeds and total area under the crop (s) for the year 1974-75, 1975-76 and 1976-77.

(xiii) Agricultural productivity is roughly measured by the gross value of agricultural output (at 1961 prices) per hectare. The output of 10 major crops, grown in Bihar, covering about 85 per cent of the gross sown area are included for this purpose.

(xiv) The following items of infrastructural facilities related to agriculture are used for preparing the composite index of infrastructure for the districts (1) irrigation facilities roughly measured by the percentage of gross area sown irrigated, (2) credit facilities i.e. - availability of agricultural credit through various agencies per rural family, (3) rural literacy rate in percentage, (4) percentage of village electrified, (5) road length per thousand square kilometres of area, (6) number of regulated markets per block and (7) number of composite demonstrations conducted for paddy and wheat in the district per block (see Appendix II (b) for method used for preparing the indices).

(xv) Only three items of new agricultural practices viz. - use of machines, use of chemical fertilisers and use of HYV seeds are included in preparing the adoption index.

(xvi) State figure refers to the average income of only 18343 persons in 2925 Sample families from 12 districts of Bihar. Regional figures are computed by taking the weighted mean of the estimated per capita income of the relevant districts in the region.

(xvii) Regional averages are computed by taking the simple mean of the scores of relevant districts in the region.

(3) Tribal concentration in the Chotanagpur region is borne out by the fact that about 92.64 per cent of the total tribals of the state reside in this region forming 32.12 per cent of the total population of the region as a whole as per 1971 census. Scheduled caste peoples are also equally (socially) backward and so they can be included in measuring the magnitude of socially backward people in the region. It is found that the total percentage of Scheduled Tribe and Scheduled Caste population in the region is highest (42.05%) in Chotanagpur followed by South Bihar Plain (18.65%) and North Bihar Plain (14.70%) as per the 1971 census. Socio-economic attitude of the tribals should therefore be given their due emphasis in the economic analysis of this region. A brief sketch of the same as brought out in some previous studies is given below. We restrict ourselves to the attitudes relevant for our study.

The Socio-economic attitude of the tribals, as noted by Mohsin and Sinha [85] are those of village communities still living in a "pre-capitalistic" or "pre-monetary" system. Land is the only productive resource. The tribals rightly flaunt their attachment to the land, but neither do they protect it from intruders, nor do they treat it well. While other economic attitudes impede their progress, their attitude towards landed property imperils their very survival [8 p.17]. Their attitude discourage the emergence of strongly featured personalities in the community or rugged individuals keen to go their own way. Poverty, then, is an exigency which makes one submit the management of one's property to social pressures, and, as such, is a truly tribal economic attitude [8 p.31].

Generally, tribal farmers do not desire and also do not have any surplus income or wealth over and above a bare minimum due to their limited knowledge of the 'possible', limited tastes, limited inquisitiveness and their natural conservatism. Most of the tribals are still primitive in their life style and lack calculating ability. Due to illiteracy they may not be aware of new agricultural practices. They may not also respond to prices inspite of their desire, unless they are given some financial assistance because initial capital base as well as the finance needed for additional investment is lacking. Thus the farmers are caught in the tangle of vicious-circles, social, economic, agricultural and health and educational, which are further bound into what is known as the Gardian knots [84].

It should be noted that the tribals of Chotanagpur significantly differ from other Indian tribals in many respects such as the following : (a) Most of the tribals of Chotanagpur are settled agriculturists and possess land. There is evidence that many of them (32.5% as per 1961 census) possess more than 5 acres of cultivable land [49]. (b) The tribal farming system is almost the same as that of the non-tribals in the region. Despite the abundance of hills and forests in the region, the tribals of Chotanagpur prefer to live in villages. They also do not practise shifting cultivation with the exception of a few minor tribal communities like the Hill Kharia of Singhbhum and the Souria Paharia of Santhal Pargana [139]. As per the information available for the year 1960-61, we find that only 5.9 per cent of the total tribal population engaged in cultivation used shifting cultivation in Bihar as against 75.5 per cent, 66.6 per cent and 59.7 per cent in

Manipur, Orissa and Tripura [110]. (c) The system of communal ownership of land among the tribals, except among the Ho tribals, is not found in Chotanagpur but is observed by some scholars in the case of other Indian states in their respective studies [52,114]. (d) Tribal integration with the Hindu society is, of course, a notable feature of Chotanagpur tribals. "In Chotanagpur in most areas, we have numerous Hindu communities living in the tribal village for many centuries" [116 p.223].

In fact all these special characteristics of Chotanagpur tribals which distinguish them from other Indian tribals are mainly attributable to the rapid social change observed in Chotanagpur over the last few decades or so. Studies indicating social change among the tribals of Chotanagpur are by Lall [75], Prasad [104], Sachidananda [116] and Vidyarthi [140].

It is pertinent to note that social changes among the tribals have not been attained equally among all tribals since different tribes are found to be at different levels of culture. The Oraon, for instance, lead a far more settled existence than do the Birhors who are still fairly nomadic in their ways [104 p. 332]. "The Santals are by far the best cultivators of their land .... the plains of the areas where they live give a clear indication of their workmanship and their knowledge of agricultural practices" [104 p.77]. Ho tribals appear to be the most rigid in their attitudes towards religion and faith and thus less affected by the alien elements. The communal attitudes are of course the special characteristics of the Ho tribals in Chotanagpur. The Hos are divided into a number of exogamous clans called killis which take their names from some animals, plant or material objects. All the members of a killi are united by a bond of kinship. In all activities the interest of the killi is primary and those of the family are subordinated to that of the killi. These characteristics have, however, undergone changes in the process of transition. Prasad has noted that "Today the family has become the dominant element in their social organisation, that is, the interests of the family are regarded as of greater importance than that of Hatu (village) or even a killi (clan). Yet these changes have not completely disintegrated the clan discipline and social solidarity. For the Ho family may still live jointly and worship the family ancesters and gods under the same roof" [104 p.108].

(4) The plateau region is full of hills and forests. About 78.7 per cent of the total area under forest in the state goes to Chotanagpur region which covers about 27.8 per cent of the total area. As a result a relatively small area is available for cultivation. Out of the available area, suitable for cultivation, the utilisation is lowest in the Chotanagpur region (56.41%) because most of the cultivable area remains idle, which may be so because of the unfavourable monsoon, want of finance, decline in fertility or socio-economic attitudes of the tribals may also be responsible for this. The highest utilisation of cultivable land is made in North Bihar Plain region followed by South Bihar Plain region with 82.84 per cent and 79.94 per cent respectively. So far as cropping intensity is concerned it is lowest in the Chotanagpur region (1.12) as compared to the North Bihar Plain (1.44) and South Bihar Plain (1.41) region.

(5) The importance of irrigation in rain-fed agriculture (Chotanagpur region has been identified as rainfed farming area of Bihar [85]) is self evident. The necessity of irrigation has considerably increased after the introduction of new technology in agriculture but the Chotanagpur region is lacking in this basic input very much. The percentage of gross area irrigated to gross area sown is only 9.35 per cent during 1975-76 in Chotanagpur region as against 23.46 per cent, 54.77 per cent and 29.76 per cent in North Bihar Plain region, South Bihar Plain region and Bihar state respectively. Besides regional disparities in the existing level of irrigation facilities, the major sources of irrigation in Chotanagpur (like tank, well and other sources) are mostly rainfed and get dried up during rabi season, unlike canals and tube wells (the major sources of irrigation in Bihar Plain regions) which are more dependable sources, providing secured irrigation facilities to the Bihar Plain regions. All this has led to a cropping pattern which is biased towards the kharif season in Chotanagpur region. The proportion of gross area sown planted during rabi season is very small in the Chotanagpur region (only 13.8%) as compared to the North Bihar Plain (41.4%) and South Bihar Plain (49.8%) regions.

(6) Transport plays an important role in the agricultural development of a region through providing good markets for inputs and outputs of the farmer. But the total length of road is found to be only 390.79 kms. per thousand square kms. of area in Chotanagpur region

as against 796.14 kms., 476.84 kms. and 540.80 kms. in North Bihar Plain, South Bihar Plain and Bihar state respectively. Other facilities which directly relate to the agricultural performance of a region are the availability of agricultural credit, extension services, and marketing facilities but these facilities are very poor in Chotanagpur region as shown in item 18, 20 and 21 in Table 2.1.

(7) Availability of infrastructural facilities significantly determines agricultural growth [142]. Unfortunately these facilities are lacking very much in the Chotanagpur region. Average indices of infrastructure for the three regions turn out to be 50.51, 92.54 and 143.88 for the Chotanagpur region, North Bihar Plain region and South Bihar Plain region respectively [19][1].

(8) Adoption of new agricultural practices is very poor in the Chotanagpur region despite a relatively higher average size of operational holdings (see item 15 and 28 in Table 2.1 for detail). This may be so because the farmers though having sufficient land are lacking in capital which is essential for purchasing new seeds, chemical fertilisers and machines. The low level of infrastructural facilities in the region may also be attributable to the low level of adoption of new agricultural practices. In our study [17] it has been observed that irrigation and credit facilities in the districts significantly explain the inter-district variation in the level of adoption of new agricultural practices in Bihar. The coefficients come out to be significant at 0.1 per cent level of significance. But unfortunately these facilities are checking in the Chotanagpur region resulting in a lower level of adoption of new agricultural practices.

It should be noted that special assistance in the form of assis tance in removing the acidity characteristics of the soils[2] is needed as a pre-requisite for adoption of new agricultural practices for the

---

1. Besides revealing the existence of inter-regional differences in the level of infrastructural facilities in Bihar, this study sufficiently indicates the direct and indirect role of infrastructure in determining the level of agricultural productivity in Bihar districts.

2. Acidity is generally measured by pH value, soil with pH less than 6.5 are acidic whereas pH more than 7.4 are alkalic and pH 6.6 to 7.3 are neutral in reaction.

farmers of Chotanagpur region because the acid soils not only give low yields of crops without manuring but also show poor response to fertiliser [80]. The problem of acidity is particularly severe in the Chotanagpur region. Chotanagpur farmers also need managerial assistance for proper use of new agricultural practices. The impact of acidity characteristics of the soil and poor management of the farmers on crop yield is sufficiently indicated by the experimental results conducted by the Operation Research Unit of the Birsa Agricultural University [97, 98]. Some of the experimental results are given in Table 2.2. It should also be noted that though the level of adoption of the new agricultural practices in the Chotanagpur region is very low as compared to other regions of Bihar there is sufficient evidence at the village level to indicate that the farmers, including tribal farmers, are gradually adopting the practices and sharing the benefits of green revolution [8, 16, 76, 130, 144].

TABLE 2.2: **Showing experimental results under dry-seeding conducted by the operation research unit of the Birsa Agricultural University, Ranchi during the Kharif season of 1978-79 and 1979-80**

| Item / Year | 1973-79 (Four-plot technique) | | 1979-80 (Two-plot technique) | |
|---|---|---|---|---|
| 1. Experiment Number | 1 | 3 | 1 | 5 |
| 2. Village (district) | Barhu (Ranchi) (Ranchi) | Bara-semartoli | Barwatoli (Ranchi) | Nowatoli (Ranchi) |
| 3. Crop (Variety) local) | Paddy (Bala & & local) | Ragi (A-404 | Groundnut (AK-12-24) | Soyabean (Pb - 1) |
| 4. Area (in acre) | 0.48 | 0.40 | 0.20 | 0.28 |
| 5. Yield rate (Qtl./ Ha.) under different treatments (four-plot techniques) | | | | |
| (i) I V I M | 17.6 | 14.4 | | |
| (ii) I V F M | 6.9 | 1.8 | | |
| (iii) F V I M | 15.7 | 12.6 | | |
| (iv) F V F M | 5.1 | 1.5 | | |

*Contd.* Table 2.2

| | | | | |
|---|---|---|---|---|
| 6. Yield rate (Qtl./Ha.) under different treatments (two-plot technique) | | | | |
| (i) Limed | .. | .. | 25.20 | 16.25 |
| (ii) Unlimed | .. | .. | 19.20 | 8.50 |

Where :

IVIM = Improved Variety and Improved Management.
IVFM = Improved Variety and Farmers' Management.
FVIM = Farmer's Variety and Improved Management.
FVFM = Farmer's Variety and Farmers' Management.

Sources [97, 98]

(9) Agricultural productivity as measured by the gross value of agricultural output (at 1961 prices) per hectare during 1976-77 is also very low in the Chotanagpur region. In one of our study we observed that the lower level of agricultural productivity in the Chotanagpur region may be owing to the lower level of adoption of new agricultural practices leading to stagnation and backwardness of the region [20].[3] It should be noted, however, that the average agricultural productivity in the Chotanagpur region as compared to other regions of Bihar is not so low as observed in the case of average adoption index. This may be taken to indicate that the Chotanagpur farmers, who mostly grow traditional varieties seeds, using organic manures and traditional agricultural practices (see item 23 to 25 in Table 2.1) are quite efficient in their own traditional method of production.

(10) Agricultural income (gross value of agricultural output at constant prices) per capita rural population during 1976-77 has been used to measure the relative agricultural development of the region

---

3. This study also attempts to identify the real cause of the low level of adoption in the Chotanagpur region in the light of the a priori hypotheses and concluded that the low level of adoption in the Chotanagpur region does not, however, mean that the farmers are irresponsive to the new agricultural practices but they do not get an adequate opportunity to respond in the sense that the viable minimum of infrastructure needed for the successful use of the new agricultural practices does not exist. The package of new technology is also not appropriate in the sense that it does not suit the pockets of the poor farmers, especially tribals.

in Bihar. It shows that the Chotanagpur region is the most backward among them having the lowest per capita agricultural income (see item 29 in Table 2.1). The average per captia annual income (at current prices) from all sources in rural areas during 1976-77 as computed from the study conducted by the Directorate of Statistics and Evaluation, Patna [ 44 ] indicates a relatively lower value for Chotanagpur (Rs. 732.16) as compared to North Bihar Plain (Rs. 779.52) and South Bihar Plain (Rs. 774.81) regions. The indices of general economic development prepared on the basis of the study conducted by the Directorate of Statistics and Evaluation [43] for the year 1974-75, however, gives a slightly different result. According to this study Chotanagpur is placed next to South Bihar Plain and the lowest rank is given to the North Bihar Plain region (see item 31 in Table 2.1). One should note that the relatively higher score for the Chotanagpur region compared to the North Bihar Plain region has been the result mainly of the inclusion of some items which relate to the urban facilities mostly available in the industrial towns of Dhanbad and Singhbhum districts, but, in fact, these urban facilities do not parcolate to the indigenous population living in rural areas.

Thus the above analysis sufficiently indicates the social and economic backwardness of the Chotanagpur region as compared to other regions of Bihar. Inter-regional disparities in Bihar may be due to different soil composition and socio-economic attitudes of the people. However, Singh [129] has noted that imbalances in investment patterns since the beginning of the first five year plan have led to the neglect of the human capital in Tribal Bihar.

## 2.4 SOCIO-ECONOMIC AND AGRO-INDUSTRIAL CHARACTERISTICS OF THE DISTRICTS IN THE REGION

Even though the homogeneity of land is assured in the Chotanagpur region, the districts within the region differ among themselves in many respects. Table 2.3, showing inter-district differences within the Chotanagpur region, reveals the following fact.

(1) The districts significantly differ in size and population. Ranchi is the biggest district and Dhanbad the smallest in size. The rate of growth of population between 1961 and 1971 was highest in Palamau

(2.1% per annum) followed by Dhanbad (2.09%), Hazaribagh (2.06%), Ranchi (1.81%), Santhal Pargana (1.60%) and Singhbhum (1.59) district. All the districts are thinly populated except Dhanbad where the density of population is more than the state average. Proportion of urban population is highest in Dhanbad district where about 44 per cent people reside in urban towns followed by Singhbhum (26%), Ranchi (14%), Hazaribagh (13%), Santhal- Pargana (6%) and Palamau (5%) as per the 1971 census. The general literacy rate is positively associated with the level of urbanisation (for detail see item 4 in Table 2.3).

(2) The region is characterised as Highland Bihar with significant mining segments and promising manufacturing belt of India [ 132 ]. Minerals are distributed in all districts but mainly concentrated in Singhbhum and Hazaribagh. Industries are solely concentrated in Singhbhum and Dhanbad. Palamau and Santhal Pargana districts have been lacking in this regard. Analysis of 1971 census data reveals that all the districts with the exception of Dhanbad are primarily agricultural. The percentage of workers engaged in secondary sector is highest in Dhanbad indicating the highest level of industrialisation (see item 5 in Table 2.3).

TABLE 2.3: **Showing inter-district differences within the Chotanagpur region (Some selected indicators)**

| Characteristics/Districts | Santhal-Pargana | Hazari-bagh | Palamau | Ranchi | Singhbhum | Dhanbad | Bihar State |
|---|---|---|---|---|---|---|---|
| 1 | 2 | 3 | 4 | 5 | 6 | 7 | 8 |
| A. General Characteristics: | | | | | | | |
| 1. Total geographical area by villge paper in thousand hectares [ 42 ] | 1422 | 1798 | 1273 | 1822 | 1357 | 299 | 17330 |
| 2. (a) Total population in lakhs as per 1971 Census [ 42 ] | 31.87 | 30.20 | 15.04 | 26.11 | 24.38 | 14.66 | 563.53 |
| (b) Percentage increase in total population in 1971 Over 1961 | 16.06 | 20.65 | 21.04 | 18.11 | 15.91 | 20.99 | 17.56 |

*Contd.* Table 2.3

| 1 | 2 | 3 | 4 | 5 | 6 | 7 | 8 |
|---|---|---|---|---|---|---|---|
| 3. Rural-urban ratio as per 1971 Census [42] | 94:6 | 87:13 | 95:5 | 86:14 | 74:26 | 56:44 | 90:10 |
| 4. Literacy rate in percentage as per 1971 Census [ 42 ] | | | | | | | |
| (i) Rural literacy rate | 13.93 | 12.84 | 13.62 | 18.03 | 15.77 | 20.11 | 17.17 |
| (ii) Urban literacy rate | 48.63 | 40.68 | 45.61 | 56.14 | 44.84 | 41.72 | 44.92 |
| (iii) General literacy rate | 15.93 | 16.42 | 15.12 | 23.24 | 25.86 | 29.51 | 19.94 |
| 5. Percentage distribution of working population as per 1971 Census [42] | | | | | | | |
| (i) Primary Sector | 87.48 | 76.82 | 89.78 | 80.56 | 69.88 | 35.71 | 83.22 |
| (ii) Secondary Sector | 4.81 | 12.48 | 3.43 | 7.59 | 16.71 | 40.28 | 6.55 |
| (iii) Tertiary Sector | 7.71 | 10.70 | 6.79 | 11.85 | 13.41 | 24.01 | 10.23 |
| 6. Percentage of socially backward people as per 1971 Census [ 42 ] | | | | | | | |
| (i) Scheduled Tribe (ST) | 36.22 | 10.99 | 19.09 | 58.08 | 46.12 | 10.61 | 8.75 |
| (ii) Scheduled Caste (SC) | 7.19 | 12.07 | 25.44 | 4.83 | 3.61 | 15.18 | 14.11 |
| (iii) Total of ST & SC | 43.41 | 23.06 | 44.53 | 62.91 | 49.73 | 25.79 | 22.86 |
| B. <u>Land Utilisation</u> : | | | | | | | |
| 7. Area under forest in thousand hectares during 1975-76 [ 42 ] | 144 (10.13) | 41 (41.21) | 558 (43.83) | 424 (23.27) | 328 (24.17) | 25 (8.36) | 2821 (16.28) |
| 1 | 2 | 3 | 4 | 5 | 6 | 7 | 8 |
| 8. Land not available for cultivation (in thousand hectares) during 1975-76 [42] | 337 (23.70) | 999 (55.56) | 680 (53.42) | 625 (34.30) | 561 (41.34) | 119 (39.80) | 5514 (31.82) |
| 9. Culturable waste land (CWL) in thousand hectares during 1975-76 [ 42 ] | 441 (31.01) | 397 (22.08) | 281 (22.07) | 430 (23.60) | 392 (28.89) | 86 (28.76) | 3351 (19.34) |
| 10. Net area sown (NAS) in thousand hectares during 1975-76 [ 42 ] | 644 (45.29) | 401 (22.30) | 312 (24.51) | 767 (42.10) | 404 (29.77) | 95 (31.77) | 8465 (48.85) |

*Contd.* Table 2.3

| 1 | 2 | 3 | 4 | 5 | 6 | 7 | 8 |
|---|---|---|---|---|---|---|---|
| 11. Gross area sown (GAS) in thousand hectares during 1975-76 [ 42 ] | 703 | 454 | 370 | 869 | 442 | 104 | 11287 |
| 12. Percentage utilisation of the cultivable land during 1975-76 [ 42 ] | 59.35 | 50.25 | 52.61 | 64.08 | 50.75 | 52.89 | 71.64 |
| 13. Cropping intensity (CI) during 1975-76 | 1.09 | 1.13 | 1.19 | 1.13 | 1.09 | 1.09 | 1.33 |
| C. Land Concentration : | | | | | | | |
| 14. Concentration ratio during 1976-77 [ 113 ] | 0.605 | 0.575 | 0.621 | 0.609 | 0.589 | 0.568 | 0.574 |
| 15. Average size of operational holdings (in hectares) during 1976-77 [113] | 2.00 | 1.66 | 1.90 | 2.50 | 1.50 | 1.40 | 1.10 |
| D. Irrigation Facilities : | | | | | | | |
| 16. Percentage of total area sown irrigated during 1975-76 [ 42 ] | 8.96 | 8.15 | 20.54 | 7.36 | 6.11 | 7.69 | 29.76 |
| 17. Source-wise distribution of total area irrigated during 1975-76 in thousand hectares [ 42 ] | | | | | | | |
| (i) Canals | 10 | 3 | 8 | 10 | 10 | .. | 1180 |
| (%) | (15.87) | (8.11) | (10.52) | (15.62) | (37.04) | | (35.13) |
| (ii) Tanks | 20 | 7 | 9 | 4 | 9 | 3 | 136 |
| (%) | (31.75) | (18.92) | (11.84) | (6.25) | (33.33) | (37.50) | (4.05) |
| (iii) Tube-wells | 2 | .. | 5 | 4 | .. | .. | 1053 |
| (%) | (3.17) | | (6.58) | (6.25) | | | (31.35) |
| (iv) Other Wells | 12 | 24 | 15 | 32 | 2 | 2 | 272 |
| (%) | (19.05) | (64.86) | (19.74) | (50.00) | (7.41) | (25.00) | (8.10) |
| (v) Other Sources | 19 | 3 | 39 | 14 | 6 | 3 | 718 |
| (%) | (30.16) | (8.11) | (51.32) | (21.88) | (22.22) | (37.50) | (21.37) |
| Total (Gross) area irrigated from all sources | 63 | 37 | 76 | 64 | 27 | 8 | 3359 |
| | (100.00) | (100.00) | (100.00) | (100.00) | (100.00) | (100.00) | (100.00) |

*Contd.* Table 2.3

| 1 | 2 | 3 | 4 | 5 | 6 | 7 | 8 |
|---|---|---|---|---|---|---|---|
| E. Some Other Allied Facilities : | | | | | | | |
| 18. Availability of agricultural credit through various agencies (viz. Government, Co-operative banks & commercial banks) per rural family (in Rs.) during 1976-77 [ 37,111 ] | 43.29 | 9.65 | 15.09 | 32.17 | 17.84 | 10.81 | 46.27 |
| 19. Length of (total) road in Kms. per thousand square kms. of area as on March 31, 1975 [ 42 ] | 317.5 | 386.3 | 346.3 | 436.4 | 357.2 | 486.9 | 542.6 |
| 20. Number of composite demonstration for paddy and wheat per block during 1976-77 [ 37,41 ] | 9.44 | 17.54 | 6.44 | 10.21 | 14.22 | 6.5 | 16.54 |
| 21. Number of regulated markets per block as on October 29, 1975 [ 37 ] | 0.121 | 0.047 | 0.080 | 0.069 | 0.093 | 0.200 | 0.1618 |
| 22. Annual normal rainfall in millimetres [ 42 ] | 1377.2 | 1258.1 | 1335.1 | 1482.6 | 1434.6 | 1310.7 | 1271.9 (i) |
| F. Seasonal Cropping Pattern : | | | | | | | |
| 23. Percentage distribution of gross area sown (GAS) between different seasons during 1975-76 [ 42 ] | | | | | | | |
| (a) Kharif Season | 84.49 | 89.21 | 69.46 | 89.76 | 91.63 | 89.42 | 63.03 |
| (i) Bhadai | 13.51 | 25.33 | 32.16 | 35.44 | 10.63 | 11.54 | 18.62 |
| (ii) Aghani | 70.98 | 63.88 | 37.30 | 54.32 | 81.00 | 77.88 | 44.41 |
| (b) Rabi Season (including grama Crops and orchads | 15.51 | 10.79 | 30.54 | 10.24 | 8.37 | 10.58 | 36.97 |
| G. Use of New Inputs in Agriculture : | | | | | | | |
| 24. Total value of machines used for agricultural purposes (in Rs.) per hectare of net area sown during 1976-77 [ 37, 39 ] | 26.58 | 39.47 | 39.40 | 31.10 | 18.41 | 61.95 | 134.58 |

*Contd.* Table 2.3

| 1 | 2 | 3 | 4 | 5 | 6 | 7 | 8 |
|---|---|---|---|---|---|---|---|
| 25. Total consumption of chemical fertilisers (NPK) per hectare of net area sown (in Kg. nutrient) during 1976-77 [ 39,137 ] | 5.35 | 3.95 | 3.11 | 5.28 | 2.04 | 4.01 | 18.65 |
| 26. Percentage of area under HYV seeds to the total area under [ 39, 42 ] | | | | | | | |
| (i) Paddy | 8.76 | 6.35 | 5.19 | 9.85 | 3.14 | 2.79 | 12.71 |
| (ii) Wheat | 80.70 | 56.89 | 36.85 | 59.72 | 52.21 | 69.60 | 66.88 |
| (iii) Maize | 5.99 | 4.52 | 5.07 | 69.74 | 22.10 | 13.00 | 50.46 |
| (iv) Total of Paddy, Wheat and Maize | 12.04 | 9.27 | 8.94 | 13.43 | 4.50 | 6.70 | 28.98 |
| H. Variety-wise Area and Yield Rate of Wheat : | | | | | | | |
| 27. Percentage distribution of area (Variety-wise irrigated and un-irrigated) under wheat [ 39 ] | | | | | | | (ii) |
| HYV Wheat | | | | | | | |
| (i) Irrigated | 73.93 | 50.48 | 32.13 | 57.68 | 47.14 | 68.74 | 54.05 |
| (ii) Un-irrigated | 6.77 | 6.41 | 4.72 | 2.04 | 5.07 | 0.86 | 12.83 |
| (iii) HYV total | 80.70 | 56.89 | 36.85 | 59.72 | 52.21 | 69.60 | 66.88 |
| Traditional Variety Wheat | | | | | | | |
| (i) Irrigated | 12.98 | 28.25 | 18.43 | 30.60 | 33.45 | 29.97 | 15.08 |
| (ii) Un-irrigated | 6.32 | 14.86 | 44.72 | 9.68 | 14.34 | 0.43 | 18.04 |
| (iii) Traditional total | 19.30 | 43.11 | 63.15 | 40.28 | 47.79 | 30.40 | 33.12 |
| 28. Variety-wise (irrigated and un-irrigated) average yield rate (kg. per hectare) of wheat [ 39 ] | | | | | | | (iii) |
| HYV Wheat | | | | | | | |
| (i) Irrigated | 1506 | 1380 | 1886 | 1990 | 1469 | 1777 | 1493 |
| (ii) Un-irrigated | 924 | 821 | 868 | 819 | 906 | 886 | 963 |
| (iii) HYV total | 1457 | 1317 | 1756 | 1950 | 1414 | 1766 | 1391 |

*Contd.* Table 2.3

| 1 | 2 | 3 | 4 | 5 | 6 | 7 | 8 |
|---|---|---|---|---|---|---|---|
| Traditional Variety | | | | | | | |
| (i) Irrigated | 865 | 839 | 950 | 918 | 1083 | 747 | 1099 |
| (ii) Un-irrigated | 864 | 661 | 559 | 1015 | 399 | 950 | 781 |
| (iii) Traditional total | 865 | 778 | 674 | 942 | 878 | 750 | 926 |
| Total Wheat | 1342 | 1085 | 1072 | 1544 | 1158 | 1457 | 1237 |
| I. Miscellaneous : | | | | | | | |
| 29. Agricultural productivity (in Rs.) per hectare during 1976-77 [ 39, 42 ] | 354.57 | 344.32 | 321.87 | 302.49 | 295.78 | 250.81 | 435.52 |
| 30. Composite index of infrastructure for the year 1976-77 [ 37, 42 ] | 60.42 | 42.82 | 43.33 | 58.84 | 47.71 | 57.55 | 100.00 |
| 31. Adoption (of new agricultural practices) index for the year 1976-77 [37, 42, 137] | 40.14 | 38.80 | 35.95 | 36.45 | 23.30 | 40.37 | 100.00 |
| 32. Gross value of agricultural output (at 1961 prices) per capita rural population during 1976-77 [ 39, 41, 42 ] | 66.23 | 40.03 | 49.38 | 74.44 | 64.48 | 17.87 | 81.55 |

Supplementary notes :

(a) Figures in square brackets in the first column are the numbers allotted in the bibliography and indicate the source (s) of information for the respective items.

(b) The figures in round brackets in each of the columns 2, 3, 4, 5, 6, 7, 8 (wherever given) stand for the corresponding percentages.

*Contd.* Table 2.3

(c) Remarks on Table 2.1 applies here also for the respective items. Some additional remarks for new items are numbered in the round brackets at the end of the 8th column corresponding to the respective items and the explanation is given below.

i. Hazaribagh figure is taken from Annual Season and Crop Report 1972-73.

ii. Percentage refers to the 3 years' (1974-75 to 1976-77) average percentage computed from the actual values given in the source.

iii. Average yield rate refers to the 3 years' (1974-75 to 1976-77) average yield rate computed from the actual values given in the source.

---

(3) The tribal concentration is highest in Ranchi district having 58.08 per cent tribal population followed by Singhbhum, Santhal Pargana, Palamau, Hazaribagh and Dhanbad with 46.12 per cent, 36.22 per cent, 19.09 per cent, 10.99 per cent and 10.61 per cent tribal respectively as per 1971 census. Oraon, Kharia, and Munda are the major tribes settled in Ranchi district, Santals are settled in Santhal Pargana and Ho tribe is to be found in Singhbhum district. These five major tribes together cover about 83 per cent of the total tribal in Bihar.

It is pertinent to note here that the tribal farmers of Chotanagpur practise crop-rotation which may be looked upon as a systematic crop substitution practice. Unfortunately exact data on the extent of crop-rotation that is followed are not available. But crop-rotation practice is reported to have been followed by the Oraon, the Munda and the Santals tribals and since the Oraon tribals are mainly concentrated in Ranchi district, the Mundas in Hazaribagh and Santals in Santhal Pargana, we infer that crop-rotation system is mainly used by the farmers of Ranchi, Hazaribagh and Santhal Pargana districts.

(4) Most of the area is covered with hills and forest and therefore a relatively small proportion of total area of the districts is available for cultivation. The highest utilisation of cultivable land is attained in Ranchi district (64.08%) and the lowest in Hazaribagh (50.25%). Cropping intensity is almost the same in all the districts except in Palamau which shows a relatively higher level than other districts but

less than the state average of 1.33. Irrigation facilities are lacking in all the districts. Percentage of gross irrigated area to gross sown area is highest in Palamau district (20.27%) and lowest in Singhbhum (5.88%) as against the state average of 29.79 per cent during 1975-76. Nature is not reluctant in providing rain to this region but the distribution is very much uneven. About 90 per cent of the total rainfall occurs during the four months from June to September. The remaining 10 per cent is received during winter normally in February. A study of the rainfall figures of Bihar [23] for 27 years (1936-37 to 1962-63) shows that all the Chotanagpur districts except Palamau are heavy rain-areas. A similar pattern was obtained by Shivamaggi [126] for the period from 1960 to 1971. Most of the crops are grown during kharif season. Percentage of area under rabi crops is very small in all the districts except Palamau where about 30.54 per cent of gross sown area is under rabi crops but less than the state average of 36.97 per cent.

(5) An interesting and important feature of the Chotanagpur region is that all holdings are owned and self-operated in all districts except in Hazaribagh where about 1.31 per cent area is partly owned and partly rented [112]. The average size of operational holdings is highest in Ranchi district (2.5 hectares) and lowest in Dhanbad (1.40 hectares) as against the state average of 1.7 hectares as per the agricultural census of 1976-77. The concentration ratio of the distribution of operational holding as worked in [ 18 ] turns out to be the highest in Palamau district (0.621) during 1976-77, followed by Ranchi (0.609), Santhal Pargana (0.605), Singhbhum (0.589), Hazaribagh (0.575) and Dhanbad (0.568).

(6) Transport facilities are relatively better in Dhanbad and Ranchi districts as compared to other districts. Agricultural credit made available through various agencies is relatively higher in Santhal Pargana and Ranchi districts than others. Marketing and extension services are very poor in all the districts of Chotanagpur region.

(7) The use of HYV seeds is very much widespread in the case of wheat as compared to paddy and maize in all the districts, area under wheat, however, is a very small proportion of the total area under cultivation. The percentage of area under HYV seeds of wheat is highest in Santhal Pargana district whereas the use of HYV maize

and paddy is highest in Ranchi district. The use of other new inputs like machines and chemical fertilisers is very poor in all the districts and do not show marked differences. The wheat crop specially the HYV wheat crop is grown under irrigated conditions in all the districts except in Palamau where about half of the total area remains under rainted conditions. The yield rate of HYV wheat is much higher than the traditional variety of wheat in all the districts (see item 28 in Table 2.3). The response of irrigation is much higher in the case of HYV wheat than that of traditional variety of wheat. Unexpectedly the average yield rate of the traditional variety of wheat under un-irrigated conditions shows higher value for the districts of Ranchi and Dhanbad than that under irrigated conditions. This may be due to the allotment of better quality land for un-irrigated wheat than that of irrgated wheat.

(8) Agricultural productivity, as measured by the gross value of agricultural output per hectare, is highest in the Santhal Pargana district, where the infrastructural facilities are relatively more abundant than in other districts as indicated by the composite index of infrastructure (see item 29 & 30 in Table 2.3). The adoption index of new agricultural practices is also relatively higher for this district. The gross value of agricultural output per capita of rural population (during 1976-77) is highest in the district of Ranchi followed by Santhal Pargana, Singhbhum, Palamau, Hazaribagh and Dhanbad. It should be noted that the gross value of agricultural output per capita of rural population in 1976-77 as compared to that in 1970-71 has increased in the districts of Santhal Pargana, Ranchi and Hazaribagh whereas it has declined in Palamau, Singhbhum and Dhanbad.

The percentage of area under principal crops in the districts of Chotanagpur region along with that of the Bihar state for the year 1956-57 and 1975-76 are given in Table 2.4. The table shows that winter rice is the most important crop grown in all the districts. Next comes autumn rice in Ranchi and Singhbhum districts and maize in Santhal Pargana, Hazaribagh, Palamau and Dhanbad districts. Among the other major cereals grown during the kharif season is ragi mostly grown in Ranchi and Hazaribagh districts. Wheat and barley are the two major cereals grown during the rabi season. Gram (pulse) is also grown during the rabi season. All the cereals and millets together cover most of the area used for cultivation ranging from

63.06 per cent in Palamau to 89.47 per cent in Dhanbad during 1975-76. The area under non-food crops is very small in all the districts except Palamau where about 11.9 per cent of the gross sown area was allotted to the non-food crops during 1975-76. The proportion of area under total cereals and millets remains almost the same over the period but in the case of maize and wheat it has increased in all the cases.

All the districts of the region show almost the same cropping pattern except Palamau, which has some distinguishing features namely, (a) The percentage of area allotted to winter rice, the major kharif crop, is very low as compared to the percentage of area allotted to this crop in other districts of the region, (b) The district has been the leader in growing rabi crops, the percentage of area under wheat, barley and gram in this district being the highest among the districts. The farmers of Palamau district might have been impressed by the performance of the adjascent district of Sahabad (old) which has been considered as the most advanced district in Bihar for rabi cultivation, (c) The district also leads in producing commercial crops i.e. non-food crops.

Thus we find that though the homogeneity of soil is almost assured, the districts differ among themselves in many respects. Considering the above differences in the district characteristics with special reference to agricultural activities, industrial development, tribal concentration and geographical contiguity we can group the six districts in the following broad categories : (a) More tribal and agricultural districts - Ranchi and Singhbhum, (b) Less tribal and agricultural districts - Santhal Pargana, Hazaribagh and Palamau, and (c) Less tribal and industrial districts - Dhanbad.

## 2.5 AGRO-ECONOMIC CHARACTERISTICS OF THE CROPS GROWN IN THE REGION

These are briefly considered in this section as a knowledge of these would be helpful in understanding the supply response behaviour in respect of different crops in different districts.

TABLE 2.4 : **Showing percentage of area under principal crops in the districts of Chotanagpur region and Bihar State during 1956-57 and 1975-76**

| Crops/Districts/ | Santhal Pargana Dist. | | Hazaribagh District | | Palamau District | | Ranchi District | |
|---|---|---|---|---|---|---|---|---|
| | 1956-57 | 1975-76 | 1956-57 | 1975-76 | 1956-57 | 1975-76 | 1956-57 | 1975-76 |
| 1. Winter rice | 64.55 | 63.97 | 54.85 | 53.91 | 25.22 | 25.13 | 47.06 | 42.92 |
| 2. Autumn rice | 1.67 | 1.75 | 1.33 | 2.22 | 3.58 | 4.62 | 14.73 | 14.73 |
| 3. Total rice (Winter rice + Autumn rice + Summer rice ) | 66.23 | 65.83 | 56.18 | 56.42 | 28.79 | 29.86 | 61.82 | 57.91 |
| 4. Maize | 6.73 | 8.95 | 8.90 | 10.19 | 10.76 | 13.35 | 1.22 | 1.77 |
| 5. Ragi | 0.52 | 0.56 | 5.20 | 5.83 | 1.17 | 1.58 | 5.63 | 6.50 |
| 6. Wheat | 0.68 | 4.60 | 1.41 | 3.25 | 3.32 | 6.20 | 0.30 | 2.99 |
| 7. Barley | 0.58 | 0.72 | 0.54 | 0.52 | 4.75 | 4.48 | 0.11 | .. |
| 8. Total cereals and millets | 79.61 | 83.28 | 80.48 | 81.90 | 58.70 | 63.06 | 77.68 | 76.56 |
| 9. Gram | 1.29 | 1.57 | 0.91 | 0.77 | 11.34 | 7.60 | 0.78 | 0.68 |
| 10. Total food crops | 97.64 | 97.15 | 92.87 | 95.22 | 89.44 | 88.10 | 92.51 | 94.11 |
| 11. Total non-food crops | 2.36 | 2.85 | 7.13 | 4.78 | 10.56 | 11.90 | 7.49 | 5.89 |
| 12. Gross area sown (in thousand hectares) | 750 | 703 | 442 | 454 | 301 | 370 | 747 | 869 |

*Contd.* Table 2.4

| | Singhbhum District | | Dhanbad District | | Bihar State | |
|---|---|---|---|---|---|---|
| | 1956-57 | 1975-76 | 1956-57 | 1975-76 | 1956-57 | 1975-76 |
| 1. Winter rice | 71.63 | 75.49 | 79.46 | 72.72 | 44.58 | 40.64 |
| 2. Autumn rice | 9.73 | 5.33 | 0.32 | 0.39 | 4.65 | 5.30 |
| 3. Total rice (Winter rice + Autumn rice + Summer rice ) | 81.39 | 81.15 | 79.78 | 73.44 | 49.24 | 46.58 |
| 4. Maize | 2.29 | 2.64 | 4.35 | 6.63 | 6.04 | 7.54 |
| 5. Ragi | 0.16 | 0.22 | 1.20 | 2.59 | 1.42 | 1.61 |
| 6. Wheat | 0.12 | 1.99 | 0.05 | 4.92 | 7.52 | 16.08 |
| 7. Barley | .. | .. | .. | .. | 3.69 | 1.56 |
| 8. Total cereals and millets | 88.73 | 88.41 | 87.65 | 89.47 | 70.66 | 75.32 |
| 9. Gram | 0.67 | 0.79 | 0.05 | 0.47 | 4.89 | 2.20 |
| 10. Total food crops | 96.96 | 96.74 | 96.97 | 98.69 | 93.95 | 95.65 |
| 11. Total non-food crops | 3.04 | 3.26 | 3.03 | 1.31 | 6.05 | 4.35 |
| 12. Gross area sown (in thousand hectares) | 418 | 442 | 111 | 104 | 10119 | 11287 |

Note : Percentages are computed values.

Sources: (i) Annual Season and Crop Report 1956-57, Government of Bihar, Patna
(ii) Office of the Directorate of Statistics & Evaluation, Government of Bihar, Patna.

The cultivable land of the Chotanagpur region is distinctly divided into various toposequences mainly in two broad categories : upland and low land covering about 40 per cent and 60 per cent respectively of the total cultivable area. Generally, farmers grow autumn rice, maize, ragi etc. in upland and winter rice in low lands during the kharif season. Wheat, barley, gram etc. are grown in both upland and low land areas during the rabi season. It should be noted, however, that the water-logging problem in most of the low land areas limits its use to transplanted paddy only during the kharif season. Thus a mono-cropping system in the form of winter rice grown in low land areas (where water-logging problem exists), prevails throughout the region with a few exceptions of paira cropping[4] system in small patches of Palamau, Singhbhum and Santhal Pargana districts. Mixed cropping system (i.e. inter cropping) is also found, though in small areas, in the districts of Santhal Pargana, Hazaribagh and Palamau. It has been reported [39] that the farmers are used to grow gram with wheat or barley specially gram and barley during the rabi season.

Kharif crops are generally grown under rain-fed conditions but rabi crops, specially HYV wheat requires regulated water supply through secured irrigation facilities. However, the traditional variety wheat can also be grown under rain-fed conditions. Unlike wheat, gram and barley are mostly grown under rain-fed conditions. Table 2.5 showing the average proportion of area irrigated to the total area under the crops grown during the rabi season in the two sub-periods reveals that the irrigated area under wheat and barley has increased in all the districts. For wheat the proportion of irrigated area has shot up in all districts except Dhanbad where the increase is not spectacular because in the pre-HYV period also about 74 per cent of wheat area was under irrigation but the total area under wheat in both the periods is very low in Dhanbad. This table also indicates that irrigation facilities during rabi season have tremendously increased in the HYV period as compared to the pre-HYV period and most of the increased irrigation facilities have been utilised for growing wheat.

The method of cultivation is almost the same in all districts, consisting in the use of the traditional farm implements like wooden ploughs, wooden harrow, spade, sickle, pitch-fork etc. It is only recently that some farmers have started using tractors, pumping sets, sprayers and dusters. Generally farmers use the broadcast method in sowing autumn rice, maize, wheat, barley and gram, whereas both broadcast and transplantation methods are used in planting winter rice and ragi.

---

4. Sowing of one crop during the harvest stage of another crop is called paira cropping system. For instance, sowing lentil in the paddy field before harvesting paddy say about 15 days prior to the paddy harvest, provided, suitable soil moisture is available in the field, would be an illustration of paira cropping.

**TABLE 2.5 : Showing average proportion of area irrigated to the total area under the crops grown during the rabi season in the two sub-periods**

| District | Period | Ten years' average proportion of area irrigated to the total area sown under the crop* | | | Ten years' average irrigated area under wheat, gram and barley in thousand hectares |
|---|---|---|---|---|---|
| | | Wheat | Gram | Barley | |
| Santhal Pargana | I | 18.25 | 2.73 | 10.83 | 2.01 |
| | II | 77.91 | 2.67 | 12.90 | 12.83 |
| Hazaribagh | I | 38.98 | 8.99 | 15.14 | 2.87 |
| | II | 68.57 | 11.99 | 23.60 | 6.55 |
| Palamau | I | 12.95 | 0.73 | 2.42 | 2.18 |
| | II | 39.89 | 0.80 | 7.38 | 7.65 |
| Ranchi | I | 11.25 | 0.94 | - | 0.28[a] |
| | II | 78.02 | 1.50 | - | 7.47[a] |
| Singhbhum | I | 14.81 | 2.71 | - | 0.20[a] |
| | II | 73.56 | 1.95 | - | 2.59[a] |
| Dhanbad | I | 74.17 | - | - | 0.11[b] |
| | II | 98.25 | - | - | 1.41[b] |

Notes : I: Pre-HYV period (1956-57 to 1965-66).
II: HYV period (1967-68 to 1976-77).
a: Wheat and gram only
b: Wheat only
*: Average proportions are computed from actual values of decadal ending 1965-66 for the pre-HYV period and of decadal ending 1976-77 for the HYV period.

Winter rice and autumn rice are the two important varieties of rice traditionally grown in the region during the kharif season. Summer rice has recently been introduced by the farmers of this region after the introduction of HYV seeds in a very small proportion of land. Winter rice enjoy relatively higher status as compared to the autumn rice in the market. Winter rice is grown from sea level to an altitude of about 5000 as the maximum and it can be safely grown in the areas where rainfall exceeds 200", it is infact, the only crop that can be grown under these conditions. Where the rainfall is at least 25" in the growing season rice can be likewise grown as a rainfed crop. Both winter rice and autumn rice are sown in the months of June-July. Transplantation work of winter rice, however, continues upto 15th of August. Autumn rice is harvested in September-October whereas winter rice harvest starts in October and continues upto December. Agro-economic conditions required for growing winter rice restricts the possibility of alternative crops that can be grown on the same plot during the same season. Autumn rice, maize and ragi can, however, be considered as alternative crops because all these crops can be grown in the same plot of land and in the same (kharif) season.

It should be noted that maize and ragi requires less water than autumn rice. In fact maize cannot be grown under heavy rainfall conditions but ragi can safely be grown even under heavy rainfall conditions on slopy land. It should also be noted that ragi is mostly grown by the tribal farmers for their own consumption. It is a grain of great nutrient value and is considered more sustaining to people doing hard physical work than any other grain. It is one of the hardiest of the crops suited for dry-farming.

Wheat is the major rabi cereal grown by the farmers of all the districts. It is only recently that the farmers of Chotanagpur have given its due share to this crop except in the Palamau district where the farmers have been growing wheat traditionally. Due to the poor moisture resistance capacity of the soil, Chotanagpur farmers usually grow wheat under irrigated conditions. The best period for sowing wheat is from November 15 to November 30. Our village survey [16], however, indicates the fact that the sowing operation (specially of the tribal farmers) continues upto January which affects adversely the output of the crop. HYV wheat is very much responsive to the nitrogenous fertiliser [ 133 ] but the poor farmers of Chotanagpur do not use the prescribed dose of fertiliser resulting in a lower yield of the crop. Barley and gram can also be grown in place of wheat on the same plot of land and in the same season. It should be noted that as compared to wheat barley and gram are preferable for dry farming because these crops require less water and less care than that of wheat.

The above analysis of the agro-economic conditions of the major crops grown in the Chotanagpur region helps us among other things in determining the broad groups of substitute crops and in understanding the relevance of the rainfall and irrigation variables in supply response studies.

## 2.6 A BRIEF SUMMARY SKETCH OF THE CHOTANAGPUR REGION AS A WHOLE

Chotanagpur is one of the three natural divisions of Bihar with characteristics distinguishing it from the rest of Bihar. Chotanagpur is socially and economically backward as compared to the other regions of Bihar. Tribal concentration is a special feature of the region. However, the socio-economic attitudes of the Chotanagpur tribals differ significantly from other Indian tribals in many respects which may be the result of the rapid social change among the Chotanagpur tribals observed in the last few decades or so. Ho tribals concentrated in Singhbhum district appear to be more rigid in their cultural attitudes and less affected by the elements of social change. The practice of crop rotation is also observed among the tribal farmers specially the Oraon, Mundas and Santals.

The Chotanagpur farmers mostly grow cereals for their own consumption using their own traditional method of cultivation in which they are quite efficient. Recently they have started using new agricultural practices specially in growing wheat. The use of HYV seeds is much more widespread in the case of the wheat crop than that of paddy and maize but the proportion of area under wheat in total area sown is very small. The Chotanagpur farmers, though they have sufficient land, lack in other concomitants which are needed for modernising agriculture. There are evidences to indicate that the farmers are sufficiently motivated towards the use of the new agricultural practices but the infrastructural facilities are lacking very much in this region and this may be the cause of the non-adoption or a lower adoption of new agricultural practices and the low level of agricultural productivity observed for this region.

Though homogeneity of soil is almost assured in the region, the districts within the region differ among themselves in many respects such as the magnitude of tribal population, infrastructural facilities, adoption of new agricultural practices and agricultural productivity. Cropping pattern is almost the same in all the districts except in Palamau which possess some distinguishing characteristics. Intertemporal variation in the cropping pattern favouring maize and wheat is also indicated. Agro-economic conditions required for growing winter rice precludes the possibilities of growing other crops during the (kharif) season on the same plot of land. However, autumn rice, maize and ragi during the kharif season and wheat, barley and gram during the rabi season can be grown on the same plot of land.

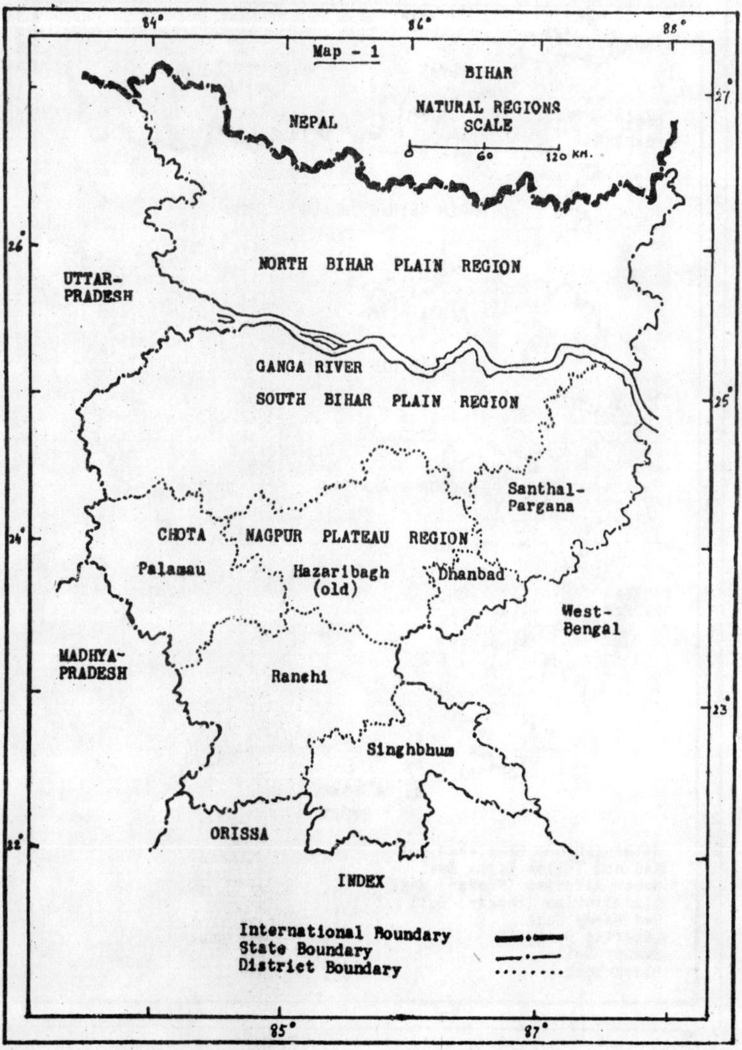
Map - 1
BIHAR
NATURAL REGIONS
SCALE
0 60 120 KM.
NEPAL
NORTH BIHAR PLAIN REGION
UTTAR-PRADESH
GANGA RIVER
SOUTH BIHAR PLAIN REGION
Santhal-Pargana
CHOTA NAGPUR PLATEAU REGION
Palamau
Hazaribagh (old)
Dhanbad
West-Bengal
MADHYA-PRADESH
Ranchi
Singhbhum
ORISSA
INDEX
International Boundary
State Boundary
District Boundary
84°
86°
88°
85°
87°
27°
26°
25°
24°
23°
22°

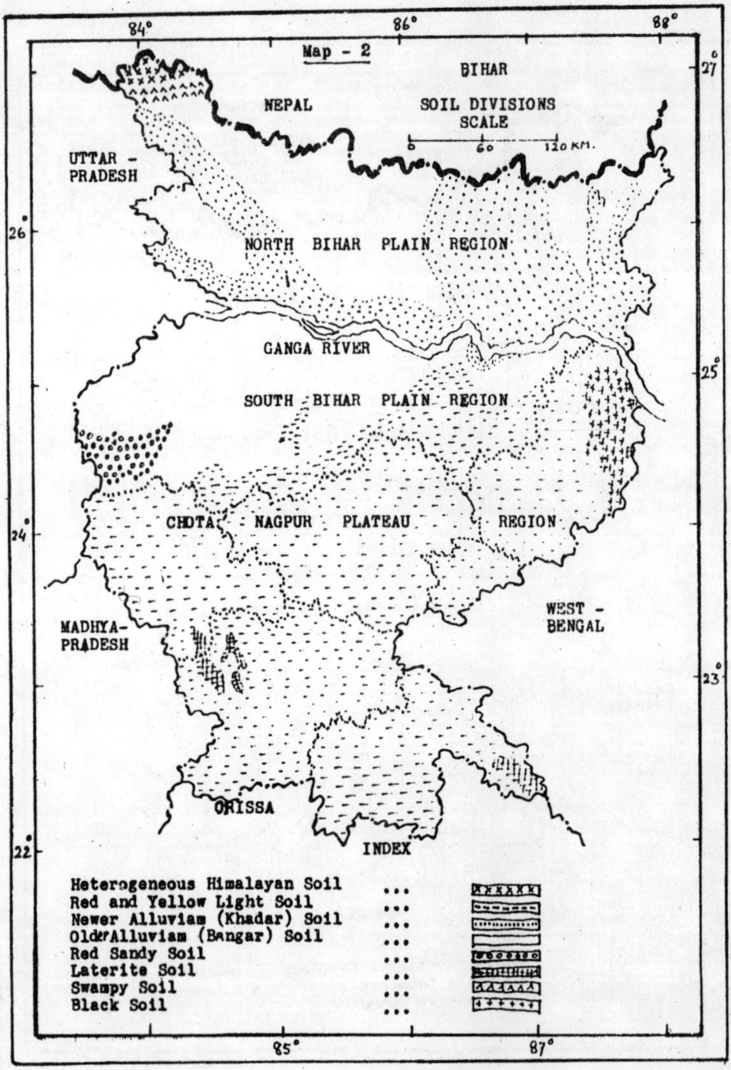
Map - 2
BIHAR
SOIL DIVISIONS
SCALE
0 60 120 KM.
NEPAL
UTTAR - PRADESH
NORTH BIHAR PLAIN REGION
GANGA RIVER
SOUTH BIHAR PLAIN REGION
CHOTA NAGPUR PLATEAU REGION
MADHYA-PRADESH
WEST - BENGAL
ORISSA
84°
86°
88°
85°
87°
27°
26°
25°
24°
23°
22°
INDEX
Heterogeneous Himalayan Soil ...
Red and Yellow Light Soil ...
Newer Alluviam (Khadar) Soil ...
Old Alluviam (Bangar) Soil ...
Red Sandy Soil ...
Laterite Soil ...
Swampy Soil ...
Black Soil ...

THREE

# BEHAVIOUR OF AGRICULTURAL OUTPUT, INPUT AND PRICES - Inter-District and Inter-Temporal Variations

## 3.1 INTRODUCTION

In this chapter we shall examine the behaviour of agricultural output, input and prices in the districts of Chotanagpur region in the two sub-periods described in the earlier chapters. Such an analysis is carried out in an attempt to provide the background profiles of the districts which would, apart from being of interest in themselves, be of great help in determining the tentative hypotheses for studying supply responses of Chotanagpur farmers. We should first look into the question of whether there has been any variation over time both within and between periods in the acreages under different crops (individually as well as in groups) in different districts and also look into the directions and the magnitudes of the variations if there are any. Then we may be able to see how much of these variations can be explained by prices and other factors.

## 3.2 METHODS USED FOR THE ANALYSIS

For this purpose we have fitted the semi-log linear trend equations of the form

$$\text{Log}_e\, Y = a + bt; \quad (t = 1, 2, ...... n)$$

For measuring agricultural output this form is more appropriate than the simple linear trend equations due to the possible multiplicative nature of the fluctuations in the figures over the years. Besides this the coefficient of t in a semi-log linear equation provides a direct estimate of the annual proportionate rate of growth. The same trend equation has also been used for estimating the trends of other items related to inputs and prices.

## 3.3 ITEMS INCLUDED IN THE DISCUSSION IN THIS CHAPTER AND RATIONALE OF SEQUENTIAL PRESENTATION

We have computed the trend equations for each relevant items (subject to the availability of time-series data) for each district of Chotanagpur region and for the two sub-periods. First of all we shall examine the trends of individual items of land-utilisation because the land-use pattern determines to a great extent the amount of land available for cultivation purposes. We shall then examine the trends of area, yield rate and output of individual crops for each district and the Chotanagpur region as a whole in the two sub-periods. The differences in the trends of area, yield-rate and output may be explained by the availability of inputs in the production process. Therefore, we shall examine the trends of different inputs like, irrigation, annual rainfall, chemical fertiliser and HYV seeds. The use of inputs specially new inputs in the HYV period may be subject to the existence of remunerative output prices. Therefore the trends of farm harvest prices have been examined and the behaviour of the prices of inputs discussed.

## 3.4 PATTERN OF LAND UTILISATION

The total land area in a region is divided into the following subcategories : (1) Forest, (2) Barren and un-culturable waste land, (3) Culturable waste land including miscellaneous tree crops and groves, (4) Permanent pasture and other grazing land, (5) Current fallow, (6) Other fallow, and (7) Net area sown. We have computed the trend equations for each type of land-use for each district and for the two sub-periods. The trend equations of gross area sown and cropping intensity have also been worked out. These are presented in Table 3.1 which exhibits the following features of the land-use pattern.

TABLE 3.1: **Showing the pattern of land utilisation in the districts of Chotanagpur region in the two sub-periods - Pre-HYV & HYV period**

(Trend equations fitted : $Log_e Y = a + bt; t = 1, 2........10$)

| Item | Santhal Pargana Pre-HYV | Santhal Pargana HYV | Hazaribagh Pre-HYV | Hazaribagh HYV | Palamau Pre-HYV | Palamau HYV |
|---|---|---|---|---|---|---|
| 1 | 2 | 3 | 4 | 5 | 6 | 7 |
| 1. Area under forest | 12.93<br>0.058*<br>(0.025)<br>0.392 | 11.87<br>-0.002<br>(0.003)<br>0.032 | 13.73<br>-0.012**<br>(0.004)<br>0.593 | 13.56<br>-0.006<br>(0.004)<br>0.208 | 13.48<br>-0.021***<br>(0.005)<br>0.703 | 13.32<br>-0.010**<br>(0.004)<br>0.449 |
| 2. Barren and unculturable land | 10.78<br>0.058***<br>(0.015)<br>0.636 | 10.50<br>-0.016***<br>(0.003)<br>0.796 | 11.42<br>0.006<br>(0.006)<br>0.116 | 11.88<br>0.015<br>(0.012)<br>0.160 | 11.10<br>-0.007<br>(0.027)<br>0.008 | 11.30<br>0.005<br>(0.004)<br>0.170 |
| 3. Land put to non-agricultural uses | 11.64<br>-0.008<br>(0.013)<br>0.047 | 11.66<br>-0.007**<br>(0.003)<br>0.342 | 11.08<br>0.040***<br>(0.008)<br>0.781 | 11.43<br>0.001<br>(0.003)<br>0.003 | 10.06<br>0.065***<br>(0.011)<br>0.825 | 10.63<br>-0.013<br>(0.008)<br>0.261 |
| 4. Culturable waste land | 11.69<br>-0.008<br>(0.033)<br>0.007 | 11.64<br>-0.008***<br>(0.001)<br>0.891 | 11.59<br>0.019**<br>(0.007)<br>0.501 | 11.65<br>-0.022***<br>(0.004)<br>0.770 | 10.93<br>-0.015<br>(0.013)<br>0.147 | 10.50<br>-0.020**<br>(0.007)<br>0.524 |
| 5. Permanent pasture and other grazing land | 10.53<br>0.055***<br>(0.016)<br>0.586 | 11.03<br>-0.003<br>(0.002)<br>0.271 | 10.65<br>0.005<br>(0.021)<br>0.008 | 10.50<br>-0.009<br>(0.008)<br>0.148 | 8.90<br>0.044**<br>(0.018)<br>0.423 | 9.32<br>-0.052***<br>(0.003)<br>0.968 |
| 6. Current fallow | 11.58<br>0.036<br>(0.026)<br>0.188 | 12.28<br>-0.019<br>(0.026)<br>0.060 | 11.90<br>-0.041*<br>(0.019)<br>0.349 | 12.05<br>0.010<br>(0.014)<br>0.061 | 11.97<br>-0.026<br>(0.021)<br>0.171 | 11.99<br>-0.004<br>(0.006)<br>0.044 |
| 7. Other fallow | 11.34<br>0.042<br>(0.027)<br>0.243 | 11.85<br>0.007<br>(0.009)<br>0.060 | 11.77<br>-0.025<br>(0.025)<br>0.113 | 11.74<br>-0.001<br>(0.004)<br>0.006 | 11.41<br>-0.-42<br>(0.042)<br>0.114 | 11.46<br>0.001<br>(0.005)<br>0.001 |
| 8. Net area sown | 13.14<br>0.003<br>(0.012)<br>0.008 | 13.19<br>0.010<br>(0.011)<br>0.085 | 12.67<br>0.028***<br>(0.006)<br>0.761 | 12.76<br>0.003<br>(0.007)<br>0.026 | 12.28<br>0.050**<br>(0.020)<br>0.435 | 12.39<br>0.026***<br>(0.006)<br>0.736 |
| 9. Gross area sown | 13.51<br>-0.007 | 13.31<br>0.009 | 13.05<br>-0.003 | 12.90<br>0.004 | 12.59<br>0.029* | 12.59<br>0.020** |

*Contd.* Table 3.1

| 1 | | 2 | 3 | 4 | 5 | 6 | 7 |
|---|---|---|---|---|---|---|---|
| | | (0.007)<br>0.111 | (0.012)<br>0.063 | (0.005)<br>0.046 | (0.007)<br>0.037 | (0.014)<br>0.361 | (0.007)<br>0.519 |
| 10. | Cropping intensity | 4.98<br>-0.011<br>(0.013)<br>0.089 | 4.71<br>0.002<br>(0.005)<br>0.025 | 4.99<br>-0.032***<br>(0.003)<br>0.916 | 4.74<br>-0.001<br>(0.006)<br>0.003 | 4.91<br>-0.021**<br>(0.009)<br>0.414 | 4.82<br>-0.009**<br>(0.003)<br>0.559 |
| | | Ranchi | | Singhbhum | | Dhanbad | |
| 1. | | 13.17<br>-0.016***<br>(0.004)<br>0.637 | 13.00<br>-0.007**<br>(0.003)<br>0.354 | 13.51<br>-0.037**<br>(0.015)<br>0.423 | 12.94<br>-0.028***<br>(0.006)<br>0.731 | 10.86<br>-0.052*<br>(0.023)<br>0.395 | 10.09<br>0.005<br>(0.003)<br>0.273 |
| 2. | | 11.44<br>0.013<br>(0.007)<br>0.287 | 11.50<br>0.010***<br>(0.002)<br>0.812 | 11.02<br>0.027*<br>(0.013)<br>0.342 | 11.60<br>0.009<br>(0.005)<br>0.299 | 10.09<br>0.042**<br>(0.016)<br>0.477 | 10.48<br>0.002<br>(0.010)<br>0.003 |
| 3. | | 11.19<br>0.014**<br>(0.005)<br>0.475 | 11.32<br>0.015***<br>(0.003)<br>0.794 | 11.31<br>0.015<br>(0.011)<br>0.181 | 11.42<br>0.030***<br>(0.004)<br>0.895 | 10.57<br>0.024<br>(0.015)<br>0.242 | 10.80<br>0.024***<br>(0.006)<br>0.652 |
| 4. | | 11.49<br>0.008<br>(0.008)<br>0.110 | 11.50<br>-0.017***<br>(0.002)<br>0.896 | 11.47<br>-0.071<br>(0.055)<br>0.171 | 11.51<br>0.002<br>(0.002)<br>0.051 | 10.66<br>-0.080<br>(0.043)<br>0.299 | 10.22<br>-0.006<br>(0.005)<br>0.161 |
| 5. | | 9.47<br>-0.021<br>(0.016)<br>0.161 | 9.13<br>-0.099***<br>(0.008)<br>0.949 | 9.72<br>-0.005<br>(0.021)<br>0.008 | 9.62<br>-0.042***<br>(0.009)<br>0.731 | 7.87<br>-0.021<br>(0.023)<br>0.095 | 7.89<br>-0.002<br>(0.015)<br>0.001 |
| 6. | | 12.10<br>-0.009<br>(0.013)<br>0.053 | 12.11<br>0.006<br>(0.005)<br>0.126 | 11.25<br>0.033**<br>(0.014)<br>0.405 | 11.68<br>0.026***<br>(0.007)<br>0.627 | 10.14<br>0.010<br>(0.018)<br>0.035 | 10.42<br>0.017<br>(0.016)<br>0.128 |
| 7. | | 12.24<br>-0.024*<br>(0.012)<br>0.329 | 12.20<br>-0.010**<br>(0.004)<br>0.368 | 11.30<br>0.022**<br>(0.008)<br>0.482 | 11.63<br>0.017***<br>(0.003)<br>0.782 | 11.14<br>-0.001<br>(0.009)<br>0.003 | 11.13<br>0.023*<br>(0.007)<br>0.564 |
| 8. | | 13.39<br>0.008<br>(0.005)<br>0.232 | 13.48<br>0.004<br>(0.003)<br>0.207 | 12.41<br>0.040**<br>(0.013)<br>0.563 | 12.88<br>0.005**<br>(0.002)<br>0.378 | 11.24<br>0.023<br>(0.020)<br>0.148 | 11.46<br>-0.022*<br>(0.011)<br>0.346 |

*Contd.* Table 3.1

| 1 | 2 | 3 | 4 | 5 | 6 | 7 |
|---|---|---|---|---|---|---|
| 9. | 13.49 | 13.54 | 12.92 | 12.93 | 11.60 | 11.50 |
| | 0.004 | 0.007 | 0.005 | 0.007** | -0.009 | -0.018 |
| | (0.004) | (0.005) | (0.004) | (0.003) | (0.006) | (0.012) |
| | 0.119 | 0.244 | 0.229 | 0.396 | 0.192 | 0.214 |
| 10. | 4.71 | 4.68 | 5.11 | 4.63 | 4.97 | 4.66 |
| | -0.004 | 0.003 | -0.035** | 0.004 | -0.032 | 0.001 |
| | (0.007) | (0.003) | (0.011) | (0.004) | (0.019) | (0.002) |
| | 0.032 | 0.068 | 0.560 | 0.114 | 0.255 | 0.007 |

**Note for this table:** Four entries in column 2, 3, 4, 5, 6, & 7 for each item stand respectively for constant, coefficient of t, its standard errors (in brackets), and $R^2$ of the regression equations fitted.

Some general notes for this table and for the tables which will appear hereafter, wherever applicable, unless and otherwise noted below the table:

(i) Pre-HYV refers to the period 1956-57 to 1965-66 and HYV refers to the period 1967-68 to 1976-77.
(ii) Figures in parentheses below the regression coefficients indicate standard errors (s.e.).
(iii) * indicates significance at 10 per cent level (two-tailed test).
** indicates significance at 5 per cent level (two-tailed test).
*** indicates significance at 1 per cent level (two-tailed test).

(1) There has been a significantly declining trend in the area under forest in all the districts during the pre-HYV period. In the HYV period also the declining trend continues in all the districts except in Dhanbad, however with a smaller rate than in the pre-HYV period. The Dhanbad district has a rising trend with a very small rate of growth. The smaller rate of decline in the second period may be due to a half-hearted afforestation policy of the government, resulting only in restricting wide-spread cutting out of the forest, whereas there might have been some real afforestation in the Dhanbad district.

(2) Barren and un-culturable land shows positive trends in the pre-HYV period in all the districts except in Palamau where it is almost stagnant. The rate of growth of barren and unculturable land is associated with the rate of deforestation. It means that most of the deforested land has gone to the category of barren and unculturable land. In the HYV period all the districts, except Santhal Pargana, show positive trends.

(3) With regard to the land put to non-agricultural use, all the districts, except Santhal Pargana, have shown rising trends in the pre-HYV period, the trend rates being highly significant in Hazaribagh, Palamau and Ranchi. In the HYV period all districts, except Santhal Pargana and Palamau, have shown rising trends, the trend rate being highly significant in Ranchi, Dhanbad and Singhbhum. Rising trends of the land used for non-agricultural purposes may be partly due to increasing urbanisation and industrialisation which adversely affects the cultivable area.

(4) Culturable waste land shows negative trend in all the districts during the pre-HYV period except in Hazaribagh and Ranchi; Hazaribagh shows significantly rising trend and in Ranchi it is almost stagnant. During the HYV period all the districts have shown negative trends except Singhbhum which shows marginally rising trend. It has been observed that the use of chemical fertilisers per hectare of net area sown is lowest in Singhbhum. The consumption of chemical fertiliser is relatively higher in the districts of Santhal Pargana, Hazaribagh and Ranchi (as noted earlier in Chapter 2) which show highly significant and negative trend coefficient of the culturable waste land. This means that the Chotanagpur farmers have utilised to some extent the culturable waste land with the use of chemical fertilisers in the HYV period.

(5) In the pre-HYV period, permanent pasture and other grazing land shows positive trends in Santhal Pargana, Hazaribagh and Palamau whereas negative trends are exhibited in the districts of Ranchi, Dhanbad and Singhbhum. During the HYV period all districts show negative trends. The rates of decline are relatively higher in the districts of Palamau, Ranchi and Singhbhum. Mass decline in the permanent pasture and other grazing land in recent years might indicate the use of such lands for growing crops for their own consumption which might have been necessitated due to rapidly growing population among the farming communities. The utilisation of such lands for cultivation might also be initiated by the availability of new inputs and also by rising prices of agricultural output. Part of the utilisation of such lands may also have been for non-agricultural purposes.

(6) So far as the fallow land, current as well as other fallow land, are concerned the districts of Santhal Pargana, Singhbhum and Dhanbad show positive trends whereas negative trends are exhibited in the districts of Hazaribagh, Palamau and Ranchi. The only district which shows a highly significant trend coefficient is Singhbhum; all other districts exhibit insignificant trends. The significantly increasing trends for Singhbhum district may be due to a general decline in the rainfall. It may also be due to a decline in the fertility of the soil.

(7) All districts show positive trends of net area sown in both the periods except Dhanbad in the HYV period. The trend of net area sown is inversely proportional to the trend of the area under forest indicating that a part of the deforested land has gone to cultivation. Increase in the net area sown may also be due to the utilisation of culturable waste land for cultivation. The negative trend in Dhanbad district during the HYV period may be the result of significantly increasing trend of the land put to non-agricultural uses and other fallow lands.

(8) Though the net area sown has been increasing, it is remarkable that the gross sown area has declined in the districts of Santhal Pargana and Hazaribagh in the pre-HYV period and in the district of Dhanbad in both the periods. Other districts have shown rising trends of gross area sown in the HYV period. It is also to be noted that the rising trend of gross area sown, where it exists, is much less than that of the net area sown. The declining trend of gross area sown and the smaller rate of its rise than that of the net area sown in the pre-HYV period are explained by the declining trend of cropping intensity in all the districts under consideration. In the HYV period, however, there has been some increase in the cropping intensity in all the districts except Hazaribagh and Palamau where the HYV programme has not made much headway.

## 3.5 BEHAVIOUR OF AREA, YIELD-RATE AND OUTPUT OF CROPS SELECTED FOR STUDYING SUPPLY RESPONSES

The examination of the direction and magnitude of change in the area, yield rate (i.e. productivity) and output of crops over time may be of crucial importance in the analysis of supply responsiveness of farmers in the region. Therefore we have computed the trend equations of area, yield-rate and output of each crops individually as well as in the aggregate for each district and Chotanagpur region as a whole for the two sub-periods. The detailed results are given in Table 3.2. Chow test has been carried out for testing the overall homogeneity of trend equations in the two sub-periods for the wheat crop only. In this case structural break is indicated by the significant F-ratios (see Table 3.3) in most cases.

TABLE 3.2: **Showing trends of area, productivity per unit of area and output of major crops in the districts of Chotanagpur region in the two sub-periods**

(Trend equations fitted : $\text{Log}_e\ Y = a + bt$)
(SANTHAL PARGANA DISTRICT)

| Crop | Area | | Productivity | | Output | |
|---|---|---|---|---|---|---|
| | Pre-HYV | HYV | Pre-HYV | HYV | Pre-HYV | HYV |
| 1 | 2 | 3 | 4 | 5 | 6 | 7 |
| 1. Winter rice | 13.04 | 12.96 | 6.86 | 6.99 | 12.98 | 13.05 |
| | -0.006 | 0.002 | 0.015 | -0.012 | 0.009 | -0.010 |
| | (0.006) | (0.011) | (0.013) | (0.015) | (0.016) | (0.022) |
| | 0.110 | 0.004 | 0.155 | 0.069 | 0.039 | 0.022 |
| 2. Autumn rice | 9.48 | 8.89 | 6.02 | 6.34 | 8.60 | 8.33 |
| | 0.001 | 0.038* | 0.034* | 0.016 | 0.035 | 0.054* |
| | (0.015) | (0.017) | (0.013) | (0.014) | (0.025) | (0.026) |
| | 0.001 | 0.389 | 0.447 | 0.141 | 0.199 | 0.353 |
| 3. Maize | 10.87 | 10.47 | 6.16 | 6.42 | 10.12 | 10.27 |
| | 0.018* | 0.028 | 0.110** | 0.022 | 0.128** | 0.050 |
| | (0.008) | (0.022) | (0.046) | (0.063) | (0.051) | (0.083) |
| | 0.399 | 0.162 | 0.413 | 0.015 | 0.442 | 0.043 |
| 4. Ragi | 8.21 | 7.78 | 5.75 | 5.78 | 7.05 | 6.66 |
| | 0.011 | 0.036 | 0.002 | 0.009 | 0.013 | 0.045 |
| | (0.013) | (0.029) | (0.059) | (0.043) | (0.066) | (0.067) |
| | 0.086 | 0.161 | 0.0002 | 0.005 | 0.005 | 0.052 |
| 5. Wheat | 8.48 | 8.44 | 5.60 | 6.63 | 7.18 | 8.16 |
| | -0.007 | 0.199*** | 0.122** | 0.059** | 0.115** | 0.258*** |
| | (0.022) | (0.018) | (0.042) | (0.022) | (0.044) | (0.025) |
| | 0.014 | 0.940 | 0.514 | 0.474 | 0.459 | 0.928 |
| 6. Barley | 8.67 | 8.63 | 5.96 | 6.24 | 7.72 | 7.96 |
| | 0.018 | -0.020 | 0.037 | 0.007 | 0.055 | -0.013 |

*Contd.* Table 3.2

| 1 | 2 | 3 | 4 | 5 | 6 | 7 |
|---|---|---|---|---|---|---|
| | (0.030)<br>0.041 | (0.014)<br>0.214 | (0.033)<br>0.142 | (0.018)<br>0.020 | (0.056)<br>0.107 | (0.024)<br>0.035 |
| 7. Gram | 9.56<br>0.014<br>(0.025)<br>0.039 | 8.56<br>0.072***<br>(0.017)<br>0.686 | 5.99<br>0.036<br>(0.023)<br>0.232 | 6.15<br>0.003<br>(0.034)<br>0.001 | 8.64<br>0.050<br>(0.038)<br>0.177 | 7.80<br>0.075**<br>(0.027)<br>0.484 |
| 8. Total Cereals (total of item 1 to 6) | 13.20<br>-0.003<br>(0.006) | 13.10<br>0.011<br>(0.012) | 6.78<br>0.023<br>(0.014) | 6.95<br>-0.010<br>(0.014) | 13.07<br>0.020<br>(0.018) | 13.14<br>0.001<br>(0.022) |
| 1 | 2 | 3 | 4 | 5 | 6 | 7 |
| | 0.037 | 0.100 | 0.256 | 0.057 | 0.131 | 0.001 |
| | | | (HAZARIBAGH DISTRICT) | | | |
| 1. Winter rice | 12.47<br>0.003<br>(0.005)<br>0.049 | 12.31<br>0.0001<br>(0.0050)<br>0.0001 | 6.74<br>0.004<br>(0.032)<br>0.002 | 6.70<br>-0.012<br>(0.023)<br>0.036 | 12.30<br>0.007<br>(0.029)<br>0.005 | 12.10<br>-0.012<br>(.028)<br>0.023 |
| 2. Autumn rice | 8.86<br>0.025<br>(0.014)<br>0.282 | 9.27<br>-0.021<br>(0.018)<br>0.144 | 5.81<br>0.028*<br>(0.015)<br>0.313 | 6.00<br>-0.009<br>(0.029)<br>0.011 | 7.76<br>0.053*<br>(0.024)<br>0.381 | 8.37<br>-0.030<br>(0.042)<br>0.060 |
| 3. Maize | 10.59<br>0.001<br>(0.006)<br>0.004 | 10.88<br>-0.020<br>(0.011)<br>0.284 | 6.29<br>0.072*<br>(0.034)<br>0.360 | 6.49<br>0.025<br>(0.053)<br>0.027 | 9.98<br>0.073*<br>(0.038)<br>0.318 | 10.46<br>0.005<br>(0.061)<br>0.001 |
| 4.Ragi | 10.09<br>0.001<br>(0.005)<br>0.001 | 10.11<br>-0.009<br>(0.020)<br>0.024 | 6.33<br>-0.011<br>(0.038)<br>0.011 | 5.87<br>0.026<br>(0.038)<br>0.055 | 9.51<br>-0.010<br>(0.037)<br>0.011 | 9.08<br>0.017<br>(0.078)<br>0.006 |
| 5. Wheat | 8.44<br>0.013<br>(0.026)<br>0.030 | 8.10<br>0.148***<br>(0.019)<br>0.800 | 5.77<br>0.086*<br>(0.039)<br>0.383 | 6.31<br>0.085**<br>(0.027)<br>0.546 | 7.30<br>0.099**<br>(0.034)<br>0.510 | 7.49<br>0.233***<br>(0.031)<br>0.875 |
| 6. Barley | 7.75<br>0.033<br>(0.038)<br>0.085 | 7.45<br>0.042**<br>(0.014)<br>0.511 | 6.08<br>0.014<br>(0.037)<br>0.018 | 6.28<br>-0.002<br>(0.031)<br>0.001 | 6.92<br>0.047<br>(0.066)<br>0.060 | 6.82<br>0.040<br>(0.030)<br>0.185 |
| 7. Gram | 8.32<br>0.014 | 7.49<br>0.084*** | 5.48<br>0.078*** | 5.98<br>0.023 | 6.99<br>0.092 | 6.56<br>0.107*** |

*Contd.* Table 3.2

| 1 | 2 | 3 | 4 | 5 | 6 | 7 |
|---|---|---|---|---|---|---|
| | (0.036) | (0.014) | (0.023) | (0.031) | (0.056) | (0.031) |
| | 0.018 | 0.815 | 0.586 | 0.064 | 0.254 | 0.592 |
| 8. Total Cereals | 12.73 | 12.66 | 6.67 | 6.63 | 12.50 | 12.38 |
| (total of | 0.003 | -0.001 | 0.010 | -0.002 | 0.013 | -0.03 |
| item 1 to 6) | (0.005) | (0.006) | (0.024) | (0.018) | (0.025) | (0.073) |
| | 0.052 | 0.001 | 0.021 | 0.002 | 0.033 | 0.002 |
| **(PALAMAU DISTRICT)** | | | | | | |
| 1. Winter rice | 11.30 | 11.37 | 6.51 | 6.11 | 10.91 | 10.57 |
| | 0.017*** | 0.002 | 0.017 | 0.067 | 0.034 | 0.069 |
| | (0.005) | (0.005) | (0.042) | (0.082) | (0.043) | (0.048) |
| 1 | 2 | 3 | 4 | 5 | 6 | 7 |
| | 0.548 | 0.033 | 0.021 | 0.199 | 0.074 | 0.198 |
| 2. Autumn rice | 9.48 | 9.39 | 5.98 | 5.97 | 8.55 | 8.45 |
| | 0.027 | 0.028*** | 0.006 | 0.011 | 0.033 | 0.039 |
| | (0.018) | (0.008) | (0.014) | (0.042) | (0.022) | (0.043) |
| | 0.212 | 0.587 | 0.022 | 0.008 | 0.213 | 0.093 |
| 3. Maize | 10.45 | 10.74 | 6.56 | 6.66 | 10.10 | 10.49 |
| | 0.005 | 0.006 | 0.007 | 0.0001 | 0.012 | 0.006 |
| | (0.007) | (0.005) | (0.048) | (0.054) | (0.052) | (0.055) |
| | 0.046 | 0.150 | 0.003 | 0.000 | 0.007 | 0.002 |
| 4. Ragi | 8.32 | 8.23 | 5.75 | 5.91 | 7.16 | 7.24 |
| | -0.002 | 0.014 | 0.044** | 0.007 | 0.042 | 0.021 |
| | (0.011) | (0.019) | (0.019) | (0.029) | (0.025) | (0.032) |
| | 0.005 | 0.066 | 0.411 | 0.007 | 0.260 | 0.050 |
| 5. Wheat | 8.78 | 9.27 | 5.77 | 6.31 | 7.64 | 8.67 |
| | 0.093** | 0.070*** | 0.086* | 0.070 | 0.179*** | 0.140** |
| | (0.033) | (0.015) | (0.038) | (0.042) | (0.038) | (0.045) |
| | 0.493 | 0.718 | 0.383 | 0.258 | 0.738 | 0.542 |
| 6. Barley | 9.40 | 9.50 | 6.08 | 6.39 | 8.61 | 8.98 |
| | 0.051 | 0.014 | 0.014 | -0.040 | 0.065 | -0.026 |
| | (0.039) | (0.023) | (0.036) | (0.027) | (0.066) | (0.034) |
| | 0.176 | 0.047 | 0.018 | 0.218 | 0.108 | 0.065 |
| 7. Gram | 10.11 | 9.32 | 5.48 | 6.14 | 8.68 | 8.56 |
| | 0.063 | 0.095** | 0.078*** | -0.030 | 0.141** | 0.065 |
| | (0.044) | (0.037) | (0.023) | (0.026) | (0.060) | (0.046) |
| | 0.204 | 0.455 | 0.585 | 0.145 | 0.409 | 0.203 |
| 8. Total Cereals | 11.92 | 12.06 | 6.50 | 6.34 | 11.51 | 11.49 |
| (total of | 0.025** | 0.012*** | 0.007 | 0.036 | 0.032 | 0.048 |
| item No.1-6) | (0.008) | (0.002) | (0.024) | (0.033) | (0.029) | (0.033) |

*Contd.* Table 3.2

| 1 | 2 | 3 | 4 | 5 | 6 | 7 |
|---|---|---|---|---|---|---|
| | 0.056 | 0.869 | 0.010 | 0.128 | 0.133 | 0.203 |
| (RANCHI DISTRICT) | | | | | | |
| 1. Winter rice | 12.75 | 12.75 | 6.58 | 6.69 | 12.43 | 12.54 |
| | 0.001 | 0.004 | 0.013 | 0.011 | 0.014 | 0.015 |
| | (0.002) | (0.003) | (0.026) | (0.016) | (0.028) | (0.016) |
| | 0.014 | 0.214 | 0.030 | 0.055 | 0.031 | 0.104 |
| 2. Autumn rice | 11.59 | 11.70 | 5.90 | 5.99 | 10.58 | 10.78 |
| | 0.0003 | 0.008*** | 0.007 | -0.004 | 0.007 | 0.004 |
| | (0.0040) | (0.002) | (0.011) | (0.018) | (0.012) | (0.022) |
| | 0.0003 | 0.772 | 0.050 | 0.006 | 0.041 | 0.005 |
| 3. Maize | 9.07 | 9.65 | 6.23 | 6.34 | 8.40 | 9.08 |
| 1 | 2 | 3 | 4 | 5 | 6 | 7 |
| | 0.008 | -0.007 | 0.070* | 0.045 | 0.078* | 0.038 |
| | (0.005) | (0.012) | (0.032) | (0.056) | (0.037) | (0.061) |
| | 0.198 | 0.043 | 0.369 | 0.075 | 0.357 | 0.047 |
| 4. Ragi | 10.65 | 10.84 | 6.14 | 6.13 | 9.89 | 10.05 |
| | 0.023 | 0.005 | -0.004 | -0.034 | 0.019 | -0.029 |
| | (0.013) | (0.008) | (0.022) | (0.042) | (0.024) | (0.047) |
| | 0.280 | 0.044 | 0.001 | 0.077 | 0.071 | 0.047 |
| 5. Wheat | 7.20 | 7.58 | 5.78 | 6.22 | 6.07 | 6.89 |
| | 0.063 | 0.238*** | 0.083* | 0.123*** | 0.146** | 0.361*** |
| | (0.049) | (0.027) | (0.039) | (0.033) | (0.048) | (0.047) |
| | 0.171 | 0.905 | 0.361 | 0.640 | 0.528 | 0.882 |
| 6. Gram | 8.39 | 6.22 | 5.48 | 6.26 | 6.96 | 7.83 |
| | 0.015 | 0.045*** | 0.078*** | -0.067* | 0.093 | -0.022 |
| | (0.033) | (0.013) | (0.023) | (0.033) | (0.052) | (0.032) |
| | 0.025 | 0.582 | 0.585 | 0.341 | 0.202 | 0.057 |
| 7. Total cereals | 13.13 | 13.18 | 6.44 | 6.50 | 12.67 | 12.77 |
| (total of | 0.003 | 0.009** | 0.011 | 0.013 | 0.014 | 0.022 |
| item 1 to 5) | (0.003) | (0.003) | (0.021) | (0.013) | (0.023) | (0.013) |
| | 0.143 | 0.500 | 0.034 | 0.122 | 0.044 | 0.253 |
| (SINGHBHUM DISTRICT) | | | | | | |
| 1. Winter rice | 12.61 | 12.70 | 6.69 | 6.79 | 12.39 | 12.58 |
| | 0.012*** | 0.004* | 0.024 | -0.003 | 0.036 | 0.001 |
| | (0.003) | (0.002) | (0.030) | (0.013) | (0.028) | (0.013) |
| | 0.620 | 0.236 | 0.077 | 0.006 | 0.168 | 0.001 |

*Contd.* Table 3.2

| 1 | 2 | 3 | 4 | 5 | 6 | 7 |
|---|---|---|---|---|---|---|
| 2. Autumn rice | 10.57 | 9.81 | 6.08 | 5.91 | 9.75 | 8.81 |
| | -0.078*** | 0.024*** | -0.025 | 0.017 | -0.103*** | 0.041** |
| | (0.011) | (0.002) | (0.020) | (0.015) | (0.015) | (0.014) |
| | 0.852 | 0.952 | 0.167 | 0.129 | 0.850 | 0.515 |
| 3. Maize | 9.34 | 9.41 | 6.99 | 7.14 | 8.53 | 9.64 |
| | 0.020 | -0.005 | 0.083** | -0.061 | 0.103*** | -0.066 |
| | (0.011) | (0.003) | (0.029) | (0.044) | (0.030) | (0.046) |
| | 0.286 | 0.207 | 0.572 | 0.191 | 0.596 | 0.203 |
| 4. Wheat | 6.15 | 6.70 | 5.77 | 6.38 | 5.02 | 6.18 |
| | 0.072 | 0.219*** | 0.086* | 0.066 | 0.158** | 0.285*** |
| | (0.041) | (0.034) | (0.038) | (0.052) | (0.057) | (0.070) |
| | 0.278 | 0.837 | 0.383 | 0.166 | 0.489 | 0.673 |
| 5. Gram | 7.72 | 8.19 | 5.48 | 6.03 | 6.29 | 7.32 |
| | 0.056** | -0.001 | 0.078*** | 0.003 | 0.134*** | 0.002 |
| | (0.024) | (0.023) | (0.023) | (0.031) | (0.038 | (0.035) |
| | 0.399 | 0.001 | 0.585 | 0.001 | 0.601 | 0.003 |
| 6. Total cereals | 12.76 | 12.79 | 6.62 | 6.77 | 12.48 | 12.65 |
| (total of | 0.006* | 0.007** | 0.027 | -0.004 | 0.033 | 0.003 |
| item 1 to 4) | (0.003) | (0.003) | (0.027) | (0.013) | (0.026) | (0.014) |
| | 0.312 | 0.410 | 0.115 | 0.014 | 0.165 | 0.003 |
| | | | (DHANBAD DISTRICT) | | | |
| 1. Winter rice | 11.38 | 11.30 | 6.98 | 7.13 | 11.46 | 11.52 |
| | 0.008 | -0.028* | 0.005 | -0.042 | -0.003 | -0.070* |
| | (0.008) | (0.012) | (0.024) | (0.028) | (0.030) | (0.037) |
| | 0.119 | 0.390 | 0.006 | 0.218 | 0.001 | 0.303 |
| 2. Maize | 8.34 | 8.83 | 6.44 | 6.66 | 7.96 | 8.59 |
| | 0.030*** | -0.008 | 0.026 | 0.001 | 0.056 | -0.007 |
| | (0.006) | (0.015) | (0.042) | (0.043) | (0.045) | (0.056) |
| | 0.719 | 0.037 | 0.046 | 0.0 | 0.159 | 0.002 |
| 3. Ragi | 7.60 | 8.04 | 5.88 | 6.53 | 6.58 | 7.66 |
| | 0.038 | -0.029 | 0.077** | -0.061 | 0.115** | -0.090 |
| | (0.023) | (0.017) | (0.025) | (0.042) | (0.035) | (0.053) |
| | 0.255 | 0.267 | 0.537 | 0.203 | 0.576 | 0.264 |
| 4. Wheat | 4.78 | 5.06 | 5.77 | 6.25 | 3.60 | 4.41 |
| | 0.022 | 0.306*** | 0.086* | 0.124** | 0.108 | 0.430*** |
| | (0.067) | (0.070) | (0.038) | (0.041) | (0.091) | (0.079) |
| | 0.013 | 0.706 | 0.384 | 0.535 | 0.156 | 0.788 |

*Contd.* Table 3.2

| 1 | 2 | 3 | 4 | 5 | 6 | 7 |
|---|---|---|---|---|---|---|
| 5. Total Cereals | 11.46 | 11.39 | 6.95 | 7.06 | 11.50 | 11.55 |
| (total of | -0.005 | -0.021 | 0.006 | -0.031 | 0.001 | -0.052 |
| item 1 to 4) | (0.006) | (0.012) | (0.023) | (0.022) | (0.028) | (0.031) |
| | 0.070 | 0.258 | 0.007 | 0.193 | 0.0 | 0.261 |
| (CHOTANAGPUR REGION AS A WHOLE) | | | | | | |
| 1. Winter rice | 14.24 | 14.21 | 6.75 | 6.82 | 14.08 | 14.12 |
| (all 6 dis- | 0.003 | 0.002 | 0.012 | -0.003 | 0.015 | -0.001 |
| tricts) | (0.003) | (0.005) | (0.038) | (0.008) | (0.024) | (0.010) |
| | 0.115 | 0.020 | 0.036 | 0.014 | 0.043 | 0.002 |
| 2. Autumn rice | 12.05 | 12.04 | 5.99 | 6.01 | 11.14 | 11.13 |
| (all 6 dis- | 0.008 | 0.011** | -0.011 | 0.001 | -0.003 | 0.012 |
| ricts except | (0.009) | (0.004) | (0.010) | (0.015) | (0.011) | (0.018) |
| Dhanbad) | 0.091 | 0.481 | 0.125 | 0.001 | 0.012 | 0.058 |
| 3. Maize | 11.93 | 12.11 | 6.34 | 6.83 | 11.36 | 12.03 |
| (all 6- | 0.011* | 0.003 | 0.068* | -0.058 | 0.079* | -0.055 |
| districts) | (0.005) | (0.011) | (0.034) | (0.054) | (0.037) | (0.059) |
| | 0.344 | 0.009 | 0.333 | 0.129 | 0.364 | 0.098 |
| 4. Ragi | 11.24 | 11.34 | 6.17 | 6.07 | 10.50 | 10.51 |
| (all 6 dis- | 0.015* | 0.002 | -0.001 | -0.015 | 0.015 | -0.013 |
| tricts | (0.008) | (0.012) | (0.018) | (0.042) | (0.022) | (0.052) |
| except | 0.295 | 0.003 | 0.0004 | 0.017 | 0.059 | 0.008 |
| Singhbhum) | | | | | | |
| 5. Wheat | 9.88 | 9.93 | 5.74 | 6.11 | 8.62 | 9.13 |
| (all 6 | 0.050 | 0.159*** | 0.093* | 0.116** | 0.143*** | 0.275*** |
| districts) | (0.028) | (0.016) | (0.039) | (0.040) | (0.038) | (0.047) |
| | 0.285 | 0.921 | 0.414 | 0.505 | 0.642 | 0.808 |
| 6. Barley | 9.95 | 9.93 | 6.08 | 6.36 | 9.13 | 9.38 |
| (only 3- | 0.040 | 0.009 | 0.015 | -0.026 | 0.055 | -0.017 |
| districts viz. | (0.033) | (0.018) | (0.033) | (0.021) | (0.058) | (0.027) |
| Santhal | 0.151 | 0.028 | 0.027 | 0.157 | 0.1010 | .049 |
| Pargana, | | | | | | |
| Hazaribagh | | | | | | |
| and Palamau) | | | | | | |
| 7. Gram | 10.71 | 9.82 | 5.70 | 6.14 | 9.51 | 9.05 |
| (only 3 dis- | 0.039 | 0.087*** | 0.057** | -0.015 | 0.096** | 0.072** |
| tricts viz.- | (0.030) | (0.025) | (0.020) | (0.022) | (0.039) | (0.030) |
| Santhal Par- | 0.180 | 0.610 | 0.502 | 0.058 | 0.432 | 0.421 |
| gana Hazaribagh | | | | | | |
| and Palamau) | | | | | | |

*Contd.* Table 3.2

| 1 | 2 | 3 | 4 | 5 | 6 | 7 |
|---|---|---|---|---|---|---|
| 8. Total Cereals (total of item 1 to 6) | 14.50<br>0.004<br>(0.003)<br>0.165 | 14.48<br>0.007<br>(0.005)<br>0.204 | 6.64<br>0.016<br>(0.019)<br>0.080 | 6.72<br>-0.008<br>(0.019)<br>0.021 | 14.23<br>0.020<br>(0.022)<br>0.093 | 14.30<br>-0.001<br>(0.020)<br>0.001 |

Note : Four entries in column 2, 3, 4, 5, 6, & 7 for each item stand respectively for constant, coefficient of t, its standard error (in brackets) and $R^2$ of the regression equations fitted.

We take up in the next section the interpretation of the results of trend equations which appeared in Table 3.2. First of all we consider the trend equations for different crops in the districts of Chotanagpur region in the two sub-periods. This is followed by the analysis of the fluctuations around the trends as measured by $1\text{-}R^2$ and changes therein over the period.

### 3.5.1. Trends of Different Crops in the Districts of Chotanagpur Region

Santhal Pargana :- Area under winter rice has marginally declined in the pre-HYV period but increased slightly in the HYV period indicating almost stagnation in the area allotted to the winter rice. Other kharif cereals like autumn rice, maize and ragi have shown positive trends in both the periods. The farmers might have managed to secure the additional land for these crops by a gradual utilisation of the culturable waste land, which is indicated by the negative trend coefficients for the culturable waste land for this district in both the periods with high significance in the HYV period. Among the rabi crops the area under wheat has marginally declined but area under gram and barley show increasing trends in the pre-HYV period. Wheat and gram show significantly increasing trends (the magnitude of growth being higher for wheat than for gram). Whereas barley has a declining trend in the HYV period. The preference for wheat in place of barley and gram during the HYV period might be explained by the availability of irrigation facilities, chemical fertilisers and HYV seeds of wheat during this period. Since winter rice dominates among the cereals the trends of area under total cereals are like those of winter rice.

Productivity of winter rice shows a positive trend, though insignificant, in the pre-HYV period but a negative trend in the HYV period and the same trend is exhibited by the average productivity of total cereals. All other kharif cereals and rabi crops show positive trends in both the periods. It should be noted that the magnitude of the productivity trend coefficient has declined in all the cases except in the case of ragi indicating the preference for extensive farming technology with more land brought under cultivation rather than for intensive farming techniques.

Output trends are positive for all the cases and in both the periods except for winter rice and barley in the HYV period. The declining trend of output in the case of winter rice is mainly the result of the decline in the productivity which may be due to the growing weather uncertainties and lack of irrigation facilities. In the case of barley, however, the declining trend of output is the result of the decline in the area under the crop which may be due to the acreage substitution of wheat for barley. It should be noted that the relative contribution of productivity in determining the growth rate of output is higher than that of area in all the cases excepting in the case of ragi in the pre-HYV period whereas the relative contribution of area is higher than that of productivity in all the cases except that of winter rice in the HYV period.

Hazaribagh :- Area under all crops show positive trends, though insignificant, in the pre-HYV period. This increase may be due to the reduction in the fallow lands and also due to the reduction in the forest areas. In the HYV period, areas under kharif cereals and total cereals remain almost stagnant as indicated by the very small and insignificant trend coefficients. But the areas under rabi crops like wheat, barley and gram show significantly increasing trends; the magnitude of the coefficient being highest for wheat followed by those for gram and barley which indicates the preference of the farmers for wheat as compared to barley and gram in the HYV period. The increase in the area under rabi crops might be the result of the rapid utilisation of culturable waste land as indicated by the highly significant negative coefficient of the culturable waste land.

Productivity shows positive trends in all the cases (highly significant for gram) in the pre-HYV period. During the HYV period productivity remains almost stagnant as indicated by the very small and insignificant trend coefficients except in the case of wheat which shows significantly positive trends.

Output trends are analogous to those of productivity trends except for barley in the HYV period which exhibits a positive output trend in spite of a declining productivity trend because of the significantly increasing trend in the area under the crop. Absolute values of trend coefficients have declined in all cases except in the case of ragi during the HYV period as compared to the pre-HYV period. The relative contribution of productivity is higher than that of area in determining the growth rate of output in all the cases and in both periods except in the case of wheat, barley and gram in the HYV period.

Palamau :- Area under all the crops show positive trends in both the periods except for ragi in the pre-HYV period. Additional area for cultivation might have come from the deforested land and the utilisation of culturable waste land including permanent pasture and other grazing land, and fallow lands. The growth rate of area is higher in the case of autumn rice, maize, ragi and gram whereas it is lower in the case of winter rice, wheat and barley in the HYV period as compared to that in pre-HYV period.

Productivity trends are positive in all the cases and in both the periods except in the case of barley and gram in the HYV period. Growth rates in the HYV period are higher as compared to those in the pre-HYV period, for winter rice and autumn rice whereas they are lower in the case of maize, ragi and wheat.

The output trends are analogous to the productivity trends except in the case of gram which shows a positive trend despite a negative productivity trend in the area under the crops, during the HYV period. It should be noted that unlike as in Santhal Pargana and Hazaribagh districts the rate of growth of output of wheat is lower in the HYV period as compared to that in the pre-HYV period. This may be due to the very low level of the adoption of the HYV technology of wheat. In all the cases the relative contribution of area is higher than that of productivity except in the case of ragi and gram in the pre-HYV period and winter rice in the HYV period.

Ranchi :- Area under winter-rice, autumn rice and total cereals remain almost stagnant in the pre-HYV period. Maize and ragi show increasing trends in the pre-HYV period whereas ragi continues to increase though at a lower rate but area under maize declines in the HYV period. Wheat and gram show positive trends in both the periods, the rates of growth being higher in the HYV period than in the pre-HYV period. It should be noted that the rate of growth of area under wheat is much higher than that of gram in the HYV period which indicates the use of culturable waste land for growing wheat and the farmers' preference for growing wheat rather than grâm. The utilisation of culturable waste land for growing wheat in the HYV period may be due to the availability of HYV seeds and irrigation facilities in the district during the period.

Productivities of winter rice and maize show positive trends in both the periods. Productivity of autumn rice marginally increased in the pre-HYV period but declined in the HYV period whereas ragi exhibits a declining trend in both the periods. Increase in area despite the falling trend in productivity for ragi might indicate the subsistence characteristics of the crop and risk-aversion attitudes of the farmers. Productivity of wheat shows a positive trend in both the periods, but gram shows a positive trend in the pre-HYV period but the trend is negative in the HYV period; this may be due to the availability of the HYV seeds of wheat and use of other new agricultural practices for growing wheat rather than gram.

Output shows positive trends in the case of all the crops in both the pèriods except in the case of ragi and gram in the HYV period. The increase in output may be due to the increase in area under good weather conditions and widespread use of HYV seeds in the HYV period. Productivity plays a relatively higher role in determining the growth rate of output in all the cases in the pre-HYV period except in the case of ragi while area dominates over productivity in the case of autumn rice, wheat and gram in the HYV period.

Singhbhum :- Acreage under winter rice, maize and total cereals show positive trends but autumn rice shows negative trend in the pre-HYV period whereas winter rice, autumn rice and total cereals show positive trends but maize a negative trend during the HYV period. Among the rabi crops wheat and gram show positive trends in the pre-HYV period whereas wheat continues to increase at a much higher rate, the area under gram declining during the HYV period indicating farmers' preference for wheat as against gram in the HYV period when irrigation facilities were made available to the farmers in the district.

Productivity of winter rice, maize and total cereals shows positive trends and autumn rice negative in the pre-HYV period whereas winter rice, maize and total cereals exhibit negative trends and autumn rice positive trend in the HYV period. Productivity of wheat and gram show positive trends in both the periods, the magnitude being lower in the HYV period than that in the pre-HYV period which may be due to the absence of HYV seeds for gram and a very small use of HYV seeds of wheat by the farmers in the district. Output trends are analogous to area trends in all the cases in both the periods. The relative role of area is greater than that of productivity in determining the growth rate of output in all the cases except that of maize and gram.

Dhanbad :- Area under winter-rice and total cereals show a negative trend but positive trends are exhibited for maize and ragi during the pre-HYV period, whereas winter rice, maize, ragi as well as the total cereals show negative trends during the HYV period. Wheat, the only rabi cereal, has shown a positive trend in both the periods, the rate of growth being higher in the HYV period than in the pre-HYV period. The decline in the cultivated area specially in the HYV period may be partly due to the use of land for non-agricultural purposes and partly due to the considerable increase in the fallow lands which may be due to weather uncertainties, decline in the fertility of the soil and also because of the non-availability of agricultural labourers due to the prevalence of a higher wage rate even for unskilled labourers in the mines and industries.

Productivities of all kharif cereals including that of total cereals show positive trends during the pre-HYV period, whereas in the HYV period the productivity of maize remains almost stagnant but those of winter rice, ragi and total cereals have declined. The decline is, however, statistically insignificant. Productivity of wheat shows a significantly positive trend in both the periods. The output trends are analogous to the area trends in all the cases. Area becomes more important than productivity in the case of winter rice and maize in the pre-HYV period and of wheat in the HYV period in determining the growth rates of output.

Chotanagpur region as a whole :- Area shows positive trends in all the cases in both the periods. However, very small and insignificant trend coefficients are observed for winter rice and total cereals. The rates of growth of area for autumn rice, maize and ragi remain almost the same in the pre-HYV period, but a significant increase (though very small in magnitude) in the area under autumn rice, maize and ragi is observed during the HYV period. Again the rates of growth of area under wheat, barley and gram, grown during the rabi season are almost the same in the pre-HYV period but the rate of growth of area under wheat is considerably higher than those of barley and gram in the HYV period, indicating a preference for wheat as compared to barley and gram which may be due to the introduction followed by a fairly widespread acceptance of HYV seeds for wheat during the period.

Productivity shows almost stagnant behaviour as indicated by the very small and insignificant trend coefficients in all the cases and in both the periods except in the case of maize and gram in the pre-HYV period and for wheat in both the periods. Output shows almost the same trend as that of productivity. Productivity is more important than area in determining the growth rate of output in the pre-HYV period and shows a great deal of fluctuation which may be due to the result of traditional farmers' dependence on monsoon in the absence of secured irrigation facilities. Area becomes more important than productivity in determining the output growth in the HYV period which might indicate that Chotanagpur farmers respond by utilising the culturable waste land and fallow land for raising the output, when irrigation and other new inputs are made available to them.

Table 3.3 showing computed F-ratios associated with the test for overall homogeneity of trend equations for wheat in the two sub-periods indicates that the acreage trend equations are significantly different in the two sub-periods in all the districts except Palamau, whereas productivity trend equations differ significantly only in Santhal Pargana and Dhanbad districts and output trend equations only in Santhal Pargana, Hazaribagh, Ranchi and Dhanbad.

TABLE 3.3: Showing Computed F-ratios associated with the test of overall homogeneity of trend equations in the two sub-periods for wheat crop (The Chow test)

Equation fitted : Loge Y = a + bt)

| District /Region | Computed F Statistics associated with the trend equations of | | |
|---|---|---|---|
| | Area (df = 2,16) | Productivity (df = 2,16) | Output (df = 2,16) |
| Santhal Pargana | 26.55*** | 13.88*** | 4.03*** |
| Hazaribagh | 10.97*** | 0.72 | 7.50*** |
| Palamau | 2.39 | 1.65 | 1.23 |
| Ranchi | 5.01** | 1.10 | 6.01** |
| Singhbhum | 3.81* | 0.27 | 1.21 |
| Dhanbad | 4.40** | 10.32*** | 3.55* |
| Chotanagpur region as a whole | 6.82** | 1.68 | 5.31** |

$$F = \frac{[RSS_p - (RSS_1 + RSS_2) / K]}{(RSS_1 + RSS_2) / (n_1 + n_2 - 2K)}$$

Where,

$RSS_p$ = Residual sum of squares in the pooled equation.

$RSS_1$ = Residual sum of squares in the equation for the pre-HYV period.

$RSS_2$ = Residual sum of squares in the equation for the HYV period.

K = Number of parameters including constant term.

$n_1$ & $n_2$ = Number of observations in the pre-HYV period and HYV period respectively.

### 3.5.2 Fluctuations Around the Trends

Fluctuations around the trend in area, productivity and output may be taken to reflect the uncertainty and risk involved in cultivation. 1-$R^2$ where $R^2$ is the multiple correlation coefficient may be taken as a measure of the fluctuation around the trend line. The values of 1-$R^2$ obtained for the regression lines estimated for the area, productivity and output of all the major cereals in chotanagpur districts indicate very high fluctuations. They also indicate that the fluctuations have increased in the HYV period in most cases except in the case of wheat. In the case of wheat the fluctuations have considerably declined in the HYV period which may be due to the increased irrigation facilities during the rabi season in the HYV period. The increased fluctuations in the HYV period in the case of other cereals may be due to the increased weather uncertainties, as most of the kharif crops are grown under rain-fed conditions and in the absence of secured irrigation facilities the farmers have to depend on monsoon rainfall.

## 3.6 BEHAVIOUR OF INPUTS USED

In this sub-section we have examined the behaviour of inputs used in the production of agricultural output in terms of the trends of important items like irrigated area, annual rainfall, use of chemical fertilisers and HYV seeds. First of all we examine the trends of irrigated area and then consider the pattern of rainfall and the trends in the use of chemical fertilisers and HYV seeds.

### 3.6.1 Irrigation

We have computed the trends of overall net irrigated area, gross irrigated area, irrigation intensity and irrigated area under major crops in the districts of Chotanagpur region in the two sub-periods. The only district that has shown a significantly rising trend in the total irrigated area (both net as well as gross) during both the periods is Ranchi which happened[1] to be the administrative headquarters of the Chotanagpur region, while all other districts have shown more or less falling trends with a few exceptions (see item 1 and 2 in Table 3.4). It should be noted that the growth rate of gross irrigated area is higher than that of net irrigated area in all the districts and in both the periods except in Ranchi in the HYV period. This fact is also revealed by the positive irrigation intensity trends in all the districts for both the periods except Ranchi in the HYV period (see item 3 in Table 3.4).

---

1. In 1977, Chotanagpur has been divided into North Chotanagpur and South Chotanagpur with headquarters at Hazaribagh and Ranchi respectively.

If we consider the distribution pattern of irrigation facilities for individual crops we find that in Bihar during 1976-77 about 83.45 per cent of the gross irrigated area goes to winter rice (44.46%) and wheat (38.99%) only. We have, therefore, fitted the trend equations for these two crops separately, which appear in Table 3.4.

TABLE 3.4: **Showing trends of irrigated area and total annual rainfall in the districts of Chotanagpur region in the two subperiods**

(Trend equation fitted : $Log_e Y = a + bt$)

| Item | Santhal Pargana Pre-HYV | Santhal Pargana HYV | Hazaribagh Pre-HYV | Hazaribagh HYV |
|---|---|---|---|---|
| 1 | 2 | 3 | 4 | 5 |
| 1. Net irrigated area (NIA) (in hectare) | 11.00<br>0.028<br>(0.068)<br>0.021 | 11.05<br>-0.034**<br>(0.014)<br>0.417 | 11.14<br>-0.132***<br>(0.013)<br>0.688 | 10.02<br>0.012<br>(0.017)<br>0.055 |
| 2. Gross irrigated area (GIA) (in hectare) | 10.96<br>0.043<br>(0.069)<br>0.045 | 11.10<br>-0.017<br>(0.015<br>0.143 | 11.06<br>-0.083***<br>(0.021)<br>0.655 | 10.06<br>0.029<br>(0.019)<br>0.225 |
| 3. Irrigation intensity i.e GIA/NIA)x100 | 4.57<br>0.012***<br>(0.003)<br>0.658 | 4.66<br>0.017***<br>(0.004)<br>0.675 | 4.52<br>0.080**<br>(0.019)<br>0.460 | 4.65<br>0.018*<br>(0.009)<br>0.312 |
| 4. Irrigated area under winter rice (in hectare) | 11.41<br>-0.015<br>(0.027)<br>0.036 | 11.25<br>-0.128***<br>(0.017)<br>0.871 | 11.06<br>-0.189***<br>(0.035)<br>0.787 | 9.50<br>-0.175***<br>(0.029)<br>0.821 |
| 5. Irrigated area under wheat (in hectare) | 6.66<br>0.011<br>(0.033)<br>0.015 | 7.42<br>0.293***<br>(0.025)<br>0.945 | 7.33<br>0.042*<br>(0.022)<br>0.306 | 7.56<br>0.166***<br>(0.034)<br>0.749 |
| 6. Total annual rainfall (in - millimetres) | 7.32<br>-0.024<br>(0.021)<br>0.140 | 7.21<br>-0.008<br>(0.024)<br>0.013 | 7.09<br>-0.006<br>(0.019)<br>0.014 | 7.06<br>0.0<br>(0.0)<br>0.0 |
| | <u>Palamau</u> | | <u>Ranchi</u> | |
| 1. | 11.13<br>0.002<br>(0.004)<br>0.028 | 11.14<br>-0.007<br>(0.007)<br>0.090 | 8.99<br>0.051<br>(0.052)<br>0.108 | 9.58<br>0.118***<br>(0.020)<br>0.818 |

*Contd.* Table 3.4

| 1 | 2 | 3 | 4 | 5 |
|---|---|---|---|---|
| 2. | 11.10 | 11.15 | 8.92 | 9.77 |
| | 0.013** | 0.001 | 0.078 | 0.107*** |
| | (0.005) | (0.006) | (0.051) | (0.022) |
| | 0.412 | 0.006 | 0.224 | 0.756 |
| 3. | 4.58 | 4.63 | 4.54 | 4.79 |
| | 0.012*** | 0.005** | 0.028*** | -0.012*** |
| | (0.002) | (0.002) | (0.008) | (0.003) |
| | 0.788 | 0.365 | 0.611 | 0.682 |
| 4. | 11.01 | 10.02 | 9.13 | 9.48 |
| | 0.009 | -0.017* | -0.089* | -0.025 |
| | (0.006) | (0.008) | (0.045) | (0.027) |
| | 0.235 | 0.375 | 0.342 | 0.099 |
| 5. | 7.02 | 8.02 | 4.31 | 6.55 |
| | 0.43 | 0.121*** | 0.180*** | 0.341*** |
| | (0.037) | (0.023) | (0.017) | (0.025) |
| | 0.139 | 0.771 | 0.932 | 0.959 |
| 6. | 7.21 | 7.15 | 7.26 | 7.23 |
| | -0.017 | -0.007 | -0.009 | -0.002 |
| | (0.016) | (0.018) | (0.014) | (0.014) |
| | 0.124 | 0.017 | 0.043 | 0.003 |
| | Singhbhum | | Dhanbad | |
| 1. | 10.34 | 9.96 | 9.14 | 8.32 |
| | -0.037 | -0.004 | -0.059 | -0.024 |
| | (0.017) | (0.012) | (0.070) | (0.060) |
| | 0.372 | 0.016 | 0.082 | 0.020 |
| 2. | 10.29 | 10.02 | 9.11 | 8.29 |
| | -0.022 | 0.003 | -0.043 | 0.001 |
| | (0.016) | (0.012) | (0.069) | (0.056) |
| | 0.200 | 0.005 | 0.047 | 0.0001 |
| 3. | 4.36 | 4.66 | 4.57 | 4.57 |
| | 0.015*** | 0.007 | 0.017* | 0.026*** |
| | (0.004) | (0.005) | (0.008) | (0.006) |
| | 0.667 | 0.203 | 0.345 | 0.693 |
| 4. | 10.15 | 9.94 | 8.87 | 8.50 |
| | -0.014 | -0.043** | -0.043 | -0.285*** |
| | (0.018) | (0.018) | (0.091) | (0.083) |
| | 0.074 | 0.430 | 0.027 | 0.593 |

Contd. Table 3.4

| 1 | 2 | 3 | 4 | 5 |
|---|---|---|---|---|
| 5. | 3.54 | 5.92 | 4.13 | 5.21 |
| | 0.209** | 0.283*** | 0.070 | 0.288*** |
| | (0.071) | (0.036) | (0.075) | (0.063) |
| | 0.552 | 0.885 | 0.099 | 0.724 |
| 6. | 7.28 | 7.15 | 7.22 | 7.30 |
| | -0.019 | -0.005 | -0.022 | -0.020 |
| | (0.015) | (0.012) | (0.015) | (0.023) |
| | 0.163 | 0.024 | 0.219 | 0.084 |

Note: Four entries in column 2, 3, 4, & 5 for each item stand respectively for constant, coefficient of t, its standard errors (in brackets), and $R^2$ of the regression equations fitted.

The trend coefficients of irrigated area under winter rice are negative for all the districts and for both the periods, except in Palamau which shows a very small positive trend in the pre-HYV period ; the decline in the HYV period is also the least in the Palamau district. Decline in the irrigated area under winter rice may be one of the important factors for explaining the declining productivity trends for the crop in most of the districts of Chotanagpur during the HYV period. On the contrary the trend cocfficients of wheat are positive for all the districts in both the periods except in Santhal Pargana in the pre-HYV period. It should be noted that the growth rate of irrigated area under wheat is much higher in the HYV period than in the pre-HYV period in all the districts. The increase in area and output of wheat in the HYV period may be attributable to the increased irrigation facilities available during the rabi season and its use for growing wheat.

### 3.6.2 **Rainfall**

We have fitted the trend equations for the annual rainfall for all districts in the two sub-periods and presented in item 6 of Table 3.4. Besides this, we have also computed the mean and coefficient of variation of the pre-sowing rainfall relevant to different cropping seasons in the two sub-periods and presented in Table 3.5. We find negative trend coefficients in all the districts in both the periods except Hazaribagh which shows stagnant behaviour. All the coefficients came out insignificant. $R^2$ values are very small and the values of $R^2$s have declined in the HYV period indicating increased weather uncertainties in the HYV period. Besides this, average annual rainfall has also declined in the HYV period in all the districts except in Ranchi, Hazaribagh and Dhanbad. However, average pre-sowing rainfall during the Bhadai season has declined and coefficient of variation increased in all the districts during the HYV period. Pre-sowing rainfall during Aghani season remains almost the same but the coefficient of variations have increased during the HYV period in most of the districts (see Table 3.5). During rabi season the pre-sowing rainfall has declined and coefficient of variations have increased in all the districts except in Dhanbad where average pre-sowing rainfall has increased and coefficient of variation has declined.

TABLE 3.5: **Showing mean and coefficient of variation of pre-sowing rainfall in the districts of Chotanagpur region in the two sub-periods**

| District | Period | $\bar{X}$ & CV | Mean ($\bar{X}$) and coefficient of variation (CV) of pre-sowing rainfall (in millimetres) relevant to the | | | |
|---|---|---|---|---|---|---|
| | | | Kharif Cereal | | Rabi Cereal (September to Novembe | Total Cereal (May to November |
| | | | Bhadai (May to July) | Aghani (June to August) | | |
| Santhal-Pargana | I | $\bar{X}$ | 574.3 | 782.7 | 431.7 | 1222.6 |
| | | CV | 0.205 | 0.118 | 0.414 | 0.221 |
| | II | $\bar{X}$ | 569.2 | 819.2 | 356.6 | 1240.4 |
| | | CV | 0.344 | 0.278 | 0.201 | 0.215 |
| Hazaribagh | I | $\bar{X}$ | 478.7 | 716.3 | 379.7 | 1115.1 |
| | | CV | 0.203 | 0.181 | 0.302 | 0.180 |
| | II | $\bar{X}$ | 492.6 | 760.1 | 320.7 | 1110.4 |
| | | CV | 0.377 | 0.283 | 0.377 | 0.209 |
| Palamau | I | $\bar{X}$ | 521.3 | 823.2 | 339.0 | 1178.8 |
| | | CV | 0.218 | 0.173 | 0.238 | 0.152 |
| | II | $\bar{X}$ | 458.8 | 780.6 | 305.7 | 1103.9 |
| | | CV | 0.261 | 0.184 | 0.403 | 0.130 |
| Ranchi | I | $\bar{X}$ | 567.0 | 879.9 | 366.9 | 1280.2 |
| | | CV | 0.146 | 0.137 | 0.304 | 0.141 |
| | II | $\bar{X}$ | 561.5 | 902.1 | 357.2 | 1303.4 |
| | | CV | 0.226 | 0.187 | 0.381 | 0.128 |
| Singhbhum | I | $\bar{X}$ | 569.1 | 811.1 | 404.0 | 1268.5 |
| | | CV | 0.183 | 0.166 | 0.241 | 0.176 |
| | II | $\bar{X}$ | 476.2 | 801.1 | 303.2 | 1159.6 |
| | | CV | 0.173 | 0.170 | 0.334 | 0.117 |

*Contd. Table 3.5*

| | | | | | | |
|---|---|---|---|---|---|---|
| Dhanbad | I | $\bar{X}$ | 502.6 | 744.7 | 379.0 | 1165.0 |
| | | CV | 0.142 | 0.102 | 0.415 | 0.194 |
| | II | $\bar{X}$ | 549.2 | 842.8 | 402.8 | 1299.5 |
| | | CV | 0.357 | 0.328 | 0.354 | 0.226 |

Notes : I = Pre-HYV period
II = HYV period

TABLE 3.6: **Showing trends and annual average consumption of chemical fertilisers (NPK) in the districts of Chotanagpur region in the two sub-periods**

(Trend equations fitted : $\log_e Y = a + bt$)

| District | Period | Constant | Coefficient of t | $R^2$ | Annual average consumption of chemical fertilisers (NPK) in thousand metric tonne |
|---|---|---|---|---|---|
| Santhal-Pargana | Pre-HYV | 5.05 | 0.410*** (0.055) | 0.948 | 0.64 |
| | HYV | 8.11 | -0.048 (0.062) | 0.069 | 2.88 |
| Hazari-bagh | Pre-HYV | 5.68 | 0.170 (0.089) | 0.548 | 0.51 |
| | HYV | 7.47 | -0.077 (0.073) | 0.123 | 1.50 |
| Palamau | Pre-HYV | 5.24 | -0.048 (0.054) | 0.213 | 0.16 |
| | HYV | 6.02 | 0.0601 (0.066) | 0.092 | 0.67 |
| Ranchi | Pre-HYV | 5.76 | 0.276*** (0.028) | 0.970 | 0.78 |
| | HYV | 8.16 | -0.027 (0.065) | 0.021 | 3.60 |
| Singhbhum | Pre-HYV | 5.02 | 0.198*** (0.041) | 0.885 | 0.29 |
| | HYV | 6.98 | -0.040 (0.073) | 0.037 | 1.10 |
| Dhanbad | Pre-HYV | 3.73 | 0.267*** (0.039) | 0.948 | 0.10 |
| | HYV | 6.07 | -0.083 (0.069) | 0.155 | 1.32 |

Notes : (i) Total consumption of chemical fertilisers is in nutrient (NPK) metric tonne.
(ii) Pre-HYV period refers to the period from 1961-62 to 1965-66.

### 3.6.3 Chemical Fertilisers

Table 3.6 showing the trend equations of total consumption of chemical fertilisers (NPK) in the districts of Chotanagpur region in the two sub-periods indicate highly significant positive trends in all the districts except in Hazaribagh and Palamau in the pre-HYV period. In the HYV period the trend coefficients are negative though insignificant, in all the districts except in Palamau which shows a positive trend. It should be noted that the average consumption of chemical fertilisers in the HYV period is much higher than that in the pre-HYV period. Therefore, negative and insignificant trend coefficients in the HYV period might indicate an insufficient supply of fertilisers in the region in relation to its effective demand, and consequently it has been used by the farmers as and when made available to them. Non-availability of fertilisers in sufficient amount might be a constraint on the adoption of HYV technology.

TABLE 3.7 **Showing trends of total area under HYV seeds in the districts of Chotanagpur region during the HYV period**

(Trend equations fitted : $Log_e = a + bt$)

| District/ Crops | Paddy (total) | Wheat | Maize | Total of Paddy, Wheat & Maize |
|---|---|---|---|---|
| 1. Santhal-Pargana | 6.75 | 5.63 | 7.63 | 8.38 |
| | 0.350*** | 0.558*** | 0.013 | 0.271*** |
| | (0.047) | (0.102) | (0.082) | (0.044) |
| | 0.873 | 0.788 | 0.003 | 0.823 |
| 2. Hazaribagh | 8.64 | 5.86 | 5.43 | 8.68 |
| | 0.176*** | 0.519*** | 0.391*** | 0.253*** |
| | (0.046) | (0.089) | (0.054) | (0.040) |
| | 0.076 | 0.850 | 0.897 | 0.868 |
| 3. Palamau | 8.53 | 6.13 | 6.37 | 8.69 |
| | 0.107*** | 0.479*** | 0.197*** | 0.210*** |
| | (0.015) | (0.109) | (0.071) | (0.037) |
| | 0.887 | 0.762 | 0.559 | 0.842 |
| 4. Ranchi | 8.45 | 6.39 | 8.36 | 9.07 |
| | 0.287*** | 0.425*** | 0.051 | 0.263*** |
| | (0.024) | (0.059) | (0.050) | (0.028) |
| | 0.945 | 0.866 | 0.113 | 0.918 |
| 5. Singhbhum | 6.53 | 5.24 | 5.59 | 7.10 |
| | 0.529*** | 0.560*** | 0.312*** | 0.498*** |
| | (0.056) | (0.132) | (0.046) | (0.057) |
| | 0.937 | 0.750 | 0.885 | 0.927 |

*Contd.* Table 3.7

| | | | | |
|---|---|---|---|---|
| 6. Dhanbad | 5.83 | 4.44 | 4.31 | 6.23 |
| | 0.455*** | 0.534*** | 0.343*** | 0.467*** |
| | (0.076) | (0.137) | (0.056) | (0.074) |
| | 0.856 | 0.717 | 0.862 | 0.869 |

Notes :(i) Four entries in each column for each item stand respectively for constant, coefficient of t, its s.e. (in brackets) and $R^2$ of the regression equations fitted.

(ii) Trend equations are fitted from the data of 1967-68 to 1976-77 for the districts of Santhal Pargana and Ranchi and from 1969-70 to 1976-77 for the districts of Hazaribagh, Palamau, Singhbhum and Dhanbad.

### 3.6.4 **HYV Seeds**

Trend equations of area under HYV seeds (absolute amount in hectares) of paddy, wheat and maize have been worked out for each district of the Chotanagpur region in the HYV period and presented in Table 3.7.

The trend coefficient of area under HYV paddy and wheat are positive and highly significant in all the districts, the rate of growth being higher in the case of wheat than in the case of paddy. The area under HYV maize shows a positive trend in all the districts but the coefficient comes out significant only in the districts of Hazaribagh, Dhanbad, Singhbhum and Palamau. Total area under all the three crops combined shows a positive and highly significant trend in all the districts. The rate of growth is relatively higher in the districts of Dhanbad and Singhbhum as compared to other districts in the region. It should be noted, however, that the use of HYV seeds is started earlier in Santhal Pargana and Ranchi as compared to other districts of Chotanagpur region.

Thus the above analysis of the input use pattern indicates the fact that the Chotanagpur farmers are sensitive to the use of irrigation facilities but these facilities are lacking to a great extent in the region. It is only recently, after the introduction of HYV technology that some irrigation facilities are made available to the farmers during the rabi season which have been mostly used for growing wheat. Increased irrigation facilities have led to the expansion of area under wheat and provided incentives to the farmers for using HYV seeds. Besides other constraints like financial, technical and informational lack of irrigation facilities during the kharif season may be responsible for the non-acceptance or lower acceptance of HYV seeds of paddy and maize. Lack of adequate supply of chemical fertilisers may also be responsible for the lower adoption of HYV technology in the Chotanagpur region.

## 3.7 BEHAVIOUR OF PRICES

The behaviour of prices of output and inputs are separately examined in the two following sub sections. First we consider the behaviour of farm harvest prices and seasonal variations in them. We also examine how far the overall price of agricultural output, as represented by a weighted average price of total cereals with outputs as weights, has changed in relation to the All Commodity Wholesale Price Index and consider its implications. We then examine the behaviour of input prices and changes therein over the period.

### 3.7.1 **Behaviour of Output Prices**

We have fitted the trend equations (semi-log linear) of farm harvest prices of individual crops as well as the average price of total cereals for each district of Chotanagpur region and for the two-sub periods and presented them in item 1 to 8 in Table 3.8. We have also fitted the trends of the relative prices of agricultural output, as represented by the price of total cereals relative to the All Commodity Wholesale Price Index, which is taken as the terms of trade measure. Empirical evidence suggests that even under static technology and with limited variable inputs, changes in the terms of trade have small, but positive effect on aggregate supply [ 57 ]. Relative price of total cereals has been worked out as the ratio of average price of total cereals and the All Commodity Wholesale Price Index in percentage form because the farmers, who may have no other substantial source of income, generally purchase other commodities of their necessities from the income earned by selling their surplus produce. Despite its limitation we have taken the All Commodity Wholesale Price Index as a deflator in this case in the absence of any other better index. This aspect has been discussed in detail in the next chapter while discussing the selection of the price variables. The results of the trend equations of relative prices of total cereals are presented in item 9 of Table 3.8.

TABLE 3.8: Showing the trends of farm harvest prices of impor tant crops and relative price of agricultural output in the districts of Chotanagpur region in the two sub-periods

(Trend equations fitted : $Log_e Y = a + bt$)

| Crop | Santhal Pargana | | Hazaribagh | |
|---|---|---|---|---|
| | Pre-HYV | HYV | Pre-HYV | HYV |
| 1 | 2 | 3 | 4 | 5 |
| 1. Winter | 3.84 | 4.08 | 3.87 | 4.00 |
| rice | -0.021 | 0.050* | -0.020 | 0.074* |
| | (0.029) | (0.026) | (0.035) | (0.032) |
| | 0.058 | 0.353 | 0.038 | 0.403 |
| 2. Autumn | 2.56 | 4.08 | 3.89 | 4.19 |
| rice | 0.044* | 0.061** | -0.034 | 0.036 |
| | (0.021) | (0.026) | (0.040) | (0.030) |
| | 0.357 | 0.409 | 0.083 | 0.152 |
| 3. Maize | 3.09 | 4.21 | 3.21 | 4.03 |
| | 0.062* | 0.053* | 0.068** | 0.087** |
| | (0.027) | (0.027) | (0.022) | (0.028) |
| | 0.387 | 0.329 | 0.539 | 0.538 |
| 4. Ragi | 3.16 | 3.79 | 3.04 | 3.77 |
| | 0.022 | 0.071** | 0.067* | 0.111** |
| | (0.027) | (0.029) | (0.030) | (0.044) |
| | 0.078 | 0.436 | 0.391 | 0.444 |
| 5. Wheat | 3.73 | 4.53 | 3.84 | 4.43 |
| | 0.055** | 0.048* | 0.050** | 0.062** |
| | (0.020) | (0.024) | (0.015) | (0.020) |
| | 0.479 | 0.325 | 0.579 | 0.531 |
| 6. Barley | 3.52 | 4.23 | 3.48 | 4.15 |
| | 0.029 | 0.050 | 0.054* | 0.072* |
| | (0.029) | (0.030) | (0.026) | (0.035) |
| | 0.108 | 0.264 | 0.357 | 0.344 |
| 7. Gram | 3.48 | 4.36 | 3.53 | 4.35 |
| 1 | 2 | 3 | 4 | 5 |
| | 0.070*** | 0.082** | 0.082*** | 0.100*** |
| | (0.020) | (0.025) | (0.019) | (0.028) |
| | 0.608 | 0.578 | 0.702 | 0.609 |
| 8. Total | 3.80 | 4.07 | 3.78 | 3.98 |
| cereals | -0.016 | 0.058* | -0.010 | 0.082** |
| (weighted | (0.028) | (0.027) | (0.030) | (0.028) |
| average of | 0.041 | 0.395 | 0.014 | 0.508 |
| item 1 to 6) | | | | |

*Contd.* Table 3.8

| 1 | 2 | 3 | 4 | 5 |
|---|---|---|---|---|
| 9. Relative price of agricultur-ral output | 3.82<br>-0.061*<br>(0.026)<br>0.397 | 3.55<br>-0.049**<br>(0.018)<br>0.242 | 3.81<br>-0.054<br>(0.029)<br>0.301 | 3.44<br>-0.001<br>(0.024)<br>0.0002 |
| | Palamau | | Ranchi | |
| 1. | 3.75<br>-0.007<br>(0.024)<br>0.011 | 3.97<br>0.063**<br>(0.022)<br>0.502 | 3.82<br>-0.015<br>(0.026)<br>0.040 | 4.06<br>0.059**<br>(0.025)<br>0.407 |
| 2. | 3.63<br>0.013<br>(0.025)<br>0.034 | 4.02<br>0.071**<br>(0.031)<br>0.400 | 3.79<br>-0.011<br>(0.045)<br>0.008 | 4.13<br>0.057<br>(0.031)<br>0.294 |
| 3. | 2.96<br>0.087**<br>(0.021)<br>0.676 | 3.89<br>0.079*<br>(0.037)<br>0.362 | 3.42<br>0.030<br>(0.029)<br>0.116 | 4.22<br>0.052<br>(0.032)<br>0.253 |
| 4. | 3.13<br>0.030<br>(0.018)<br>0.252 | 4.02<br>0.055<br>(0.043)<br>0.173 | 3.22<br>0.048*<br>(0.024)<br>0.331 | 4.08<br>0.071**<br>(0.028)<br>0.447 |
| 5. | 3.74<br>0.048**<br>(0.017)<br>0.505 | 4.32<br>0.090**<br>(0.021)<br>0.718 | 3.67<br>0.073***<br>(0.021)<br>0.604 | 4.67<br>0.024<br>(0.025)<br>0.105 |
| 6. | 3.24<br>0.078***<br>(0.013)<br>0.811 | 3.97<br>0.094**<br>(0.035)<br>0.505 | -<br>-<br>-<br>- | -<br>-<br>-<br>- |
| 7. | 3.43<br>0.082***<br>(0.016)<br>0.769 | 4.33<br>0.088**<br>(0.033)<br>0.508 | 3.35<br>0.108***<br>(0.016)<br>0.844 | 4.64<br>0.056**<br>(0.017)<br>0.561 |
| 1 | 2 | 3 | 4 | 5 |
| 8. | 3.49<br>0.030<br>(0.019)<br>0.231 | 3.94<br>0.087**<br>(0.030)<br>0.540 | 3.78<br>-0.009<br>(0.027)<br>0.014 | 4.08<br>0.061**<br>(0.026)<br>0.409 |

*Contd.* Table 3.8

| 1 | 2 | 3 | 4 | 5 |
|---|---|---|---|---|
| 9. | 3.54 | 3.42 | 3.81 | 3.55 |
| | -0.018 | 0.002 | -0.055* | -0.022 |
| | (0.018) | (0.022) | (0.025) | (0.018) |
| | 0.118 | 0.002 | 0.378 | 0.161 |
| | Singhbhum | | Dhanbad | |
| 1. | 3.93 | 4.19 | 3.81 | 3.84 |
| | -0.021 | 0.048** | -0.018 | 0.098*** |
| | (0.023) | (0.019) | (0.025) | (0.026) |
| | 0.094 | 0.434 | 0.057 | 0.638 |
| 2. | 3.95 | 4.18 | - | - |
| | -0.019 | 0.073* | - | - |
| | (0.026) | (0.036) | - | - |
| | 0.064 | 0.436 | - | - |
| 3. | 3.42 | 4.41 | 3.27 | 4.21 |
| | 0.058* | 0.045 | 0.077*** | 0.071** |
| | (0.029) | (0.031) | (0.018) | (0.028) |
| | 0.325 | 0.211 | 0.682 | 0.448 |
| 4. | - | - | 2.93 | 3.91 |
| | - | - | 0.085*** | 0.079** |
| | - | - | (0.020) | (0.032) |
| | - | - | 0.699 | 0.425 |
| 5. | 3.79 | 4.63 | 3.70 | 4.66 |
| | 0.062*** | 0.051** | 0.057** | 0.050 |
| | (0.012) | (0.022) | (0.023) | (0.028) |
| | 0.766 | 0.405 | 0.432 | 0.282 |
| 6. | - | - | - | - |
| | - | - | - | - |
| | - | - | - | - |
| | - | - | - | - |
| 7. | 3.24 | 3.98 | - | - |
| | 0.025 | 0.064** | - | - |
| | (0.026) | (0.023) | - | - |
| | 0.105 | 0.481 | - | - |
| 8. | 3.92 | 4.18 | 3.79 | 3.81 |
| | -0.020 | 0.050* | -0.014 | 0.104*** |
| | (0.022) | (0.023) | (0.025) | (0.027) |
| | 0.089 | 0.378 | 0.035 | 0.645 |
| 9. | 3.96 | 3.65 | 3.82 | 3.28 |
| | -0.066** | -0.034* | -0.060** | 0.020 |

*Contd.* Table 3.8

| 1 | 2 | 3 | 4 | 5 |
|---|---|---|---|---|
| | (0.024)<br>0.564 | (0.015)<br>0.382 | (0.022)<br>0.477 | (0.019<br>0.120 |
| | | Bihar State | | |
| 1. | 3.81<br>-0.015<br>(0.026)<br>0.039 | 4.11<br>0.052*<br>(0.026)<br>0.342 | -<br>-<br>-<br>- | -<br>-<br>-<br>- |
| 2. | 3.80<br>-0.009<br>(0.010)<br>0.010 | 4.12<br>0.052<br>(0.063)<br>0.063 | -<br>-<br>-<br>- | -<br>-<br>-<br>- |

Notes :(i) Farm (Harvest) price of a commodity may be defined as the average wholesale price at which the commodity is disposed of by the producer to the trader at the village site during the specified harvest period [38].

(ii) Farm harvest (weighted average) price of total cereals relative to the Economic Advisors' All Commodity Wholesale Price Index is taken to measure the relative price of agricultural output.

(iii) Four entires in column 2, 3, 4, & 5 for each item stand respectively for constant, coefficient of t, its standard errors (in brackets), and $R^2$ of the regression equations fitted.

We find that the farm harvest prices of all the major cereals have increased in all the districts and in both the periods except in the few cases of winter rice and autumn rice in the pre-HYV period. Since winter rice is the most important cereal, the trend of the average price of total cereals follows almost closely that for the winter rice price (see Table 3.8). The price rise during the HYV period is much higher than that in the pre-HYV period. It should be noted that the farm harvest price of wheat relative to the farm harvest prices of its substitute crops like barley and gram has declined in most cases in the HYV period as indicated by higher growth rates of the farm harvest prices of barley and gram as compared to that of wheat. It should also be noted that the negative trends with insignificant coefficients in the case of winter rice and autumn rice in the districts of Chotanagpur region with a few exceptions in the pre-HYV period are not peculiar to this region alone but are influenced by the similar state-wide fluctuations of prices. The trend equations of Bihar state average price for these two crops have exhibited the same behaviour (see last two items in Table 3.8).

Seasonal fluctuations in the prices of output may be of great concern for the subsistence farmers because there is a general practice among the subsistence farmers to sell a substantial proportion of their agricultural produce just after the harvest to meet their urgent needs (which sometimes they have to do irrespective of the inadequate marketable surplus they possess) and buy the food grains after a lapse of some period when they fall short of them for their own consumption. The analysis of the monthly wholesale prices of agricultural commodities at Bihar State level [45] indicate relatively lower prices during the harvesting months as compared to the prices in other months of the year. Thus seasonal fluctuations in price put the subsistence farmers at double loss in the sense that they usually sell their produce at lower prices and purchase the same at higher prices.

Let us now examine the trends of relative price of total cereals as shown in item 9 in Table 3.8. We find that the trend coefficients have gone down sharply between the periods in all the districts contrary to expectation. The terms of trade have moved against agriculture which might have some dampening impact on supply responses. $1-R^2$ is very large in all the districts in both the periods and more so in the HYV period indicating greater uncertainties in the HYV period.

### 3.7.2 **Behaviour of Input Prices**

In the absence of time-series data on the input prices for each district and for the two sub-periods separately, we have collected some general information (instead of trend equations) on the prices of inputs at two points of time for comparison purposes. For this purpose we have concentrated on the costs of human labour and fertiliser because these two items constitute a greater proportion of the direct cost (about 60%) incurred for cultivation purposes in Bihar (for detail see Appendix III (a)).

We find that the price of urea ( a commonly used fertiliser with 46 per cent nitrogen ) and average daily wage of field workers have increased by about 108 per cent and 126 per cent respectively during

1976-77 as compared to that in 1967-68.[2] It should be noted that the gross return per hectare of paddy and wheat in Bihar state have increased only to the extent of 29.72 per cent and 105.25 per cent respectively in 1976-77 as compared to that in 1967-68. This increase in the cost of production relative to the return of output might have some adverse affect on area and output. In particular the increase in the price of chemical fertilisers relative to that of output might adversely affect the use of fertilisers, which might consequently affect the area under HYV seeds and output of the crops.

## 3.8 CONCLUDING REMARKS

The above analysis of the behaviour of agricultural output, input and prices indicates that Chotanagpur farmers are gradually utilising the culturable waste land and fallow lands. This may be due to various price and non-price incentives given to the the farmers and also may be due to the population increase among the farming communities. Chotanagpur farmers are mostly growing subsistence crops using the traditional methods of cultivation but recently they have started using new agricultural practices for growing wheat. The green revolution is limited to the wheat crop only and that is also more in terms of increase in area than in terms of the yield rate indicating the farmers' preference for extensive farming techniques rather than for intensive farming techniques. It is to be noted that the use of chemical fertilisers was there even in the pre-HYV period. However, the use of chemical fertilisers and irrigation has been made more popular in the HYV period. It is only the HYV seeds which have been introduced for the first time in the HYV period.

Lack of significant trends in the case of kharif cereals specially winter rice indicates stagnation in the agricultural sector with the fluctuations in area, productivity and output indicate dependence of the agriculture on monsoon. The increase in area under winter-rice is also practically, technically and economically non-feasible because

---

2. The price of urea refers to the statutorily controlled prices as reported in the Fertiliser statistics 1978-79 and the average daily wage of field workers refers to the daily wage of male workers engaged in ploughing during the kharif season taken from the office of the Directorate of Statistics and Evaluation, Patna.

of the agro-economic conditions required for growing winter rice and the very high cost involved in converting an upland into low land for cultivating winter rice. A rough estimate of the cost involved in converting an upland into low land is made and given in Appendix III (b). Increase in area under rabi crops specially that of wheat in the HYV period is made possible through the utilisation of culturable waste land and fallow land when irrigation facilities are made available to the farmers.

We find that the prices of agricultural commodities have increased to a great extent in both the periods except in a few cases in the pre-HYV period, but there is no commensurate increase in the area under the crops. This might indicate the fact that the farmers do not consider the price alone but their decisions are governed by some other factors like their dependence on monsoon, subsistence characteristics of the crops grown and fluctuations involved in the prices and constraints on the availability of cultivable land. Farmers may not also be responsive to price changes due to their socio-economic attitudes. Planting winter rice despite a fall in the price indicates the subsistence reservation attitudes of the farmers. Maintaining the area under ragi despite a fall in the productivity might indicate risk-aversion attitudes of the farmers because ragi is mostly grown by the tribal farmers for their own consumption. By planting ragi the farmers are assured of getting something even under drought situations and thus are used to grow it in a fraction of their land every year for this purpose.

The farmers' preference for growing wheat rather than barley and gram during the HYV period despite a decline in the relative price of wheat might indicate the greater importance of technological factors over price factors in this regard because the relative productivity of wheat as compared to those of barley and gram is much higher in the HYV period because of the use of HYV seeds and chemical fertilisers supported by increased irrigation facilities during the rabi season.

It appears that non-price stimuli backed by supplementary measures rather than market stimuli are more effective in the context of the backward subsistence agriculture. Subsistence characteristics of the crop, risk involved in growing the crop, input costs and the availability of technical knowledge rather than just price considerations are active in influencing supply response. Price may thus be one but not the sole consideration in determining supply response. Large and persisting rather than just marginal price differentials may also by themselves, be sufficient considerations for supply response. Thus we find some explanation for the observed changes in the area under the crops is the Chotanagpur region in terms of changes in some of the non-price factors. To what extent it is explained by price changes and non-price changes, is taken up for consideration in subsequent chapters.

The above discussion enables us to formulate some tentative hypotheses to be looked into in the chapters that follow for the analysis of the supply responsiveness of the Chotanagpur farmers such as:

(1) Farmers' responsiveness to price changes is limited which may be because of the various agricultural constraints operating on them.

(2) Farmers' response is not symmetrical in the sense that it is different when price increases and when price falls. When price increases the farmers will try to increase the acreages under the crop for greater return but when price falls in a particular year, they will not reduce the acreage under the crop but they will try to maintain it, at least to a minimum of it which, they think, is necessary for their survival if the crop is a subsistence one. We call this as subsistence reservation hypothesis.

(3) Non-price factors are more important than the price factor in the case of the subsistence crops.

(4) Peasant farmers are risk-averters rather than risk-takers.

(5) Response pattern remains almost the same in the two subperiods (i.e. in the pre-HYV period and in the HYV period) except in the case of wheat for which the use of HYV seeds in the HYV period has been quite marked.

# APPENDIX III (a)

## ITEM-WISE DISTRIBUTION OF COST OF PRODUCTION IN BIHAR

A study of inputs and output [25] based on data collected from 703 cultivator households living in 27 villages spread over seven districts and cultivating a total gross area of 5511 acres showed that taking all the cultivators and all the crops, the costs of cultivation per acre of gross area sown were as follows :

TABLE III (a) **Showing item-wise cost of production in Bihar during 1967-68 and their proportion in total cost of production**

| | Items | Cost per acre of gross area sown in Rs. | Percentage of total cost |
|---|---|---|---|
| 1. | Human labour | 120 | 48.39 |
| 2. | Bullock labour | 57 | 22.98 |
| 3. | Seeds | 38 | 15.32 |
| 4. | Manures & Fertilisers | 30 | 12.10 |
| 5. | Land revenue | 3 | 1.21 |
| | Total | 248 | 100.00 |

Source : [ 25 p. 104 ]

## APPENDIX III (b)

## ESTIMATED COST OF CONVERTING UPLAND INTO LOWLAND IN CHOTANAGPUR REGION

TABLE III (b) : **Showing estimated cost of converting upland (i.e. tanr land) into lowland (i.e. don land) for growing winter rice in the Chotanagpur region**

| Year | Estimated cost of conversion per hectare (in Rupees) | |
|---|---|---|
| | for non-industrial areas | for industrial areas |
| 1963 (as on 16.8.1963) | 6,135.54 | 6,781.38 |
| 1976 (as on 22.6.1976) | 14,144.04 | 17,179.52 |

Notes : (i) The above estimate is minimum in the sense that the cost computation assumes that the conversion will require excavation work to a depth of 3 feet only which may be treated as minimum depth. The actual depth required for conversion will, however, depend upon the height of the adjacent area to the paddy field.

(ii) The estimated cost in only a rough approximation to the actual cost, computed on the basis of schedule rate for different types of soil conservation work in the Chotanagpur region taken from the office of the Director, Soil Conservation, Government of Bihar, Patna.

FOUR

# ANALYTICAL FRAMEWORK AND SELECTION OF VARIABLES AND MODELS FOR STUDYING SUPPLY RESPONSES IN CHOTANAGPUR REGION

## 4.1 INTRODUCTION

Most of the previous studies in India deal with the problem of supply response of individual crops, by taking planted area as a proxy for planned output and compute acreage elasticity with respect to price as a measure of supply response. In the present study we have considered the concept of supply response in a broader sense. In fact, farmers' response to price changes as reflected in the absolute changes in the area allotted to a particular crop is but only one facet of supply responses. Farmers' response might be reflected in terms of crop intensity. It might be reflected in terms of changes in relative acreage of the crop also. It might be reflected in the seed-variety and quality chosen; it might also be reflected in the quality and quantity of other inputs used.

If one is interested in studying agricultural supply response (all crops taken together) comprehensively then not only absolute change in acreage but also the farmers response to price changes in terms of changes in cropping pattern and the cropping intensity would be of importance. It is as if a comprehensive joint objective or preference function incorporating the profit function is maximised subject to all the constraints and acreages under different crops, different varieties of crops and inputs used are determined. This optimisation could also be made to cover the uncertainty involved in various stages of the production process, and in the availability of inputs and in the product prices. The uncertainty in the availability of inputs particularly is very marked if rainfall and climatic factors are included in inputs. The farmer may not be consciously formulating a precisely quantitative decision-pattern as above but broadly surrogate considerations may be at work behind his decision. His lack of formal ability and his social attitudes on the one hand and trust in traditional agricultural patterns on the other might result in his following some simple working rules characterised by conservatism, traditionality, risk-aversion, subduced response to changes and response to relatively persistent and relatively large changes only.

Two groups of basic determinants lie at the back of these responses. These are (1) agricultural production conditions (inclusive of biological production process and land and climatic factors) and (2) socio-economic environmental and behavioural conditions inclusive of social and economic systems and norms and biological and psychological needs and patterns. As a resultant of these two groups of factors over the years the agricultural system in a region comes to have certain characteristics and so also do the supply responses come to have certain characteristics. By changing these determinants, changes in both the agricultural system and the supply responses operating as part of them could be gradually brought about as the earlier discussion on the acceptance of HYV seeds and new agricultural practices has indicated.

## 4.2 VARIOUS ASPECTS OF AGRICULTURAL SUPPLY RESPONSES

As argued in the above section we note below some of the characteristics of agricultural supply responses which we think are relevant in agriculture in the Chotanagpur region, some of which may be applicable to agriculture in general also.

(1) Aggregate Acreage Response :- As a result of a change in prices the farmers might increase the area under the crops provided additional land is available to them.

(2) Cropping Pattern Response:- In an economy where varieties of crops are grown and land is limited, response of one crop to changes in relative prices will be reflected more in terms of changes

in relative acreages rather than changes in absolute area under the crop. Inter-crop substitution might also lead to changes in the overall cropping pattern over the period. Inter-crop substitution within a cropping season is the most commonly practised form of substitution but in the long-run inter-seasonal cropping pattern might also undergo changes leading to changes in the overall cropping pattern even without inter-crop substitution within the cropping season. Besides acreage substitution between different substitutable crops, the following are the other dimension of substitutability.

(a) Substitution between the HYV and traditional varieties of the same crop is quite possible specially after the introduction of HYV technology because of the differences in their average yield rate of the two varieties as noted earlier in Chapter 2. It should be noted, however, that the substitution between HYV and traditional variety seeds is subject to the availability of necessary infrastructural facilities for growing HYV crops. The substitution between the new variety and the old variety seeds automatically leads to the substitution between new inputs and old inputs specially between chemical fertilisers and organic manures. Even within these varieties the substitution might also take place between the early variety seeds and late variety seeds.

(b) Winter rice and ragi are usually sown by broadcast method and transplantation method of sowing. Since transplanted rice and ragi give better yield than broadcasted rice and ragi under good weather conditions, the farmers might prefer the transplantation method of sowing.

(c) The practice of a crop-rotation might also lead to a systematic substitution of one crop for another in alternative years.

(3) Cropping Intensity Response :- As a result of a rise in prices the farmers might utilise their land twice or thrice and thus raise their output through raising the cropping intensity without changing the net area sown. But the cropping intensity response to price changes may be conditional on the availability of sufficient irrigation facilities and credit facilities which may be considered to be the preconditions for the adoption of the multiple cropping system.

(4) Input Use and through it Productivity Response :- If the possibility of increasing land input is difficult (sometimes it becomes impossible due to heavy cost involved in converting the waste land in to area suitable for growing a particular crop say winter rice) the farmers might prefer yield-raising intensive farming technique of production by using modern inputs like HYV seeds, chemical fertilisers and insecticides etc. rather than land using extensive farming technique. Under these circumstances the farmers' response may be reflected more markedly in terms of productivity response rather than acreage response to price changes. If the non-land inputs are in short supply the farmer may decide to allocate better quality land to a more profitable crop in order to increase production.

(5) Behavioural Response :- Supply response, in its generalised sense of the term, also includes behavioural response of the farmer. By behavioural response in agriculture we mean broad changes in the decision behaviour of the farmers affecting agricultural supply, in response to the price and other stimuli. For instance, due to frequent floods and droughts during the kharif season, affecting severely the output of cereals mainly paddy, the subsistence farmers may try to change their consumption habits from rice to bread and may take initiatives for growing more wheat and barley during the rabi season. The farmers may also change their attitude towards the traditional method of cultivation and might adopt new agricultural practices. The adoption of new agricultural practices itself is an important aspect of supply response.

All these various aspects of supply responses could not be brought in our present empirical study explicitly due to the paucity of relevant data but to the extent possible they have been incorporated and utilised in formulating the regression model and in interpreting the empirical results based on them.

It should be noted that ex-anté and ex-post price responses may be two different things. The farmers may be ex-anté responsive but constraints and shortages of various kinds may prevent them from being so. It could also be that the ex-anté responsiveness itself is not very clear. Also the two may be interacting. Although very often the two are not distinguished from the point of view of overall impact, it is desirable to make a distinction between them particularly from the point of view of devising remedial policies. The other point that may be noted at this stage is that one has to talk of different degrees of price responsiveness rather than just the two states - price responsiveness and price-indifference. In this study, though it is only the ex-post supply response that has been brought in in the various equations and studied, the distinction between the ex-anté and ex-post responses is kept in mind and brought in the discussion. In the sections that follow we take up the formulation of these models and the selection of variables for them drawing on the characteristics discussed above. First the Nerlovian model which is widely used in this context is briefly considered in the next section.

## 4.3 THE NERLOVIAN MODEL

As mentioned above the basic model used for studying the supply responses in agriculture is that of Nerlovian lagged adjustment model [92]. Most of the empirical research in Indian agriculture is based on either direct application or minor modification or further extension of the Nerlovian model. The quantitative specification of the Nerlovian supply response model as developed by Nerlove along with some comments is given in Appendix IV. As noted there, many problems of application and estimation are encountered when a model of Nerlovian type is used for annual crops. It has also been noted that it does not work well for perennial crops and livestocks. We consider below the suitability of the model in the context of agriculture in backward areas.

The Nerlovian model implicitly assumes a very high entrepreneurial ability with good knowledge of mathematics, of the farmers (which American farmers might possess for whom the model was primarily formulated) for involving themselves in mathematical calculations before reacting to any changes. The peasant farmers of an isolated region in a backward economy may not possess such an intrepreneurial ability and required level of mathematical knowledge. Still they engaged in cultivation as a major and perhaps as the only source of their livelihood. It might indicate the fact that the farmers' decision in such an economy might be governed by an informal decision making process without assigning any formal mechanism such as the one Nerlovian model assumes, or any other formal decision making processes. Nerlove himself has noted in his recent article [93] that "This discussion does not fully resolve the dynamic problem involved in studying agricultural supply response in the context of a developed economy moreover, it does not touch at all on the essential dynamics of supply response in developing economies."

Nerlove has also pointed out the problems in using the basic model in developing economies in which (specially during the transitional period) substantial changes are occurring in the availability of new variety seeds and of inputs other than land and labour, and in which major improvements in infrastructure are taking place. He has noted that "Moreover, more often than not, these changes are not reflected in "visible" prices, although in market-oriented economies major shifts in the demand for various agricultural commodities or in the supply of inputs used in agricultural production do take the form of price changes. In the supply response studies ... I tacitly assumed that "viable" prices convey all the information to which farmers find it necessary to respond. This is certainly not true even in recent times in a highly developed economy such as we have in the United States" [ 93 p. 883].

While reviewing the supply response studies using Nerlovian type model, Askari and Cummings have observed that "All the studies we have included .... make obvious the complexity of estimating responsiveness from price, acreage and other time series data, .... Responsiveness - estimates obtained directly through functional relationships would not only be cheaper ...., but more importantly, would be more accurate and more reflective of the dynamic factors in a developing society"[6 p. 343].

All this suggests that we need a basis different from the one underlying the Nerlovian model for analysing the behaviour of farmers in a backward economy like that of Chotanagpur which we have chosen for the present study. Chotanagpur as we have stressed a number of times earlier is a socially and economically backward region of a backward state in India. This point is taken up for a more detailed consideration in the next section.

## 4.4 NEED FOR SEPARATE MODEL OF SUPPLY RESPONSE

The need for a separate model arises mainly on account of the peculiar characteristics of the decision-making units and also the peculiarities of the production conditions under which the farmers take decisions. We first examine the peculiarities of the decision-making units and then summarise the conditions under which they (the decision-making units ) take decisions.

### 4.4.1 **Peculiar Characteristics of Decision-makers**

The farmers of tribal concentrated Chotanagpur region are poor and illiterate. Most of the tribals are still primitive in their life-style and being uneducated lack calculating ability as noted earlier in Chapter 2. The following are some of the observations which I cound make through frequent visits to tribal villages in Ranchi district, where the major tribal population are the Oraon, the Kharia and the Munda. Tribal people specially women can count up to only twenty which they call "kori" in the local dialects but they can keep an account of more than twenty in terms of the multiple of the number. Knots are used to add the memory, tying and untying respectively for addition and substraction of numbers. Because of ignorance they cannot count time and other things. They do not even know their approximate age nor that of their children. They sometimes relate their age to some important events like great famine or even the important events of the village. Avan Exem has also noted that there is general complaint that the tribal students do not know mathematics [8 p. 38]. It should be noted, however, that the lack of counting ability is not only among the tribals but also applicable to most of the poor non-tribals.

The tribal farmers may not be equipped with mathematical tools and they usually do not possess numeracy training also but it does not, however, necessarily mean that they are irrational. They may be quite rational in their behaviour though it may not be the quantied, accurate rationality specified to modern norms. Take an instance of a tribal person who cannot count time as noted earlier. He cannot read the exact time indicated in a watch but he has a fairly accurate idea of the time. He goes to his field and returns home at the right times. After observing the height of the sun in the sky, he can give an estimate of the time and one would be surprised that it is approximately the same as by a watch; the difference between this estimate and exact time may be insignificant. What we want to emphasize is that an innumerate farmer may not be in a position to maximise his net return in the true sense of the mathematical term according to the economic theory norms but he may be quite rational in his behaviour in a broad sense and according to a different set of norms."[1]

Schultz [119] on the other hand commends the efficiency of the traditional farmer. He argues that farmers in traditional agricultural settings, while they may differ for reasons of schooling, health and experience in their ability to perceive, interpret, and respond to new events as they impinge on their farm enterprises, do in fact allocate the limited resources at their disposal in a highly efficient manner. Mellor [83] has also noted that "Subsistence farmers in traditional economies may appear to farm inefficiently because they are judged by criteria different from the ones they use in making their decisions; that farmers in fact are operating quite efficiently given their environmental conditions and objectives the variability in decisions indicates more variability in environment and objectives than variability in efficiency." Economic rationality of the tobacco grower in Malawai (Africa) has already been established by Dean [33]

To sum up we may say that though the Chotanagpur farmers may not be supposed to keep a complicated mathematical mechanism in their mind while taking decisions about the allocation of land to different crops, an informal rational decision-making process in line with their tribal norms seems to be operating in the case of these poor farmers who are generally accused of practising merely traditional and hence sub-rational method of cultivation.

--------------------

1. Deshpande [35] has, however, noted that a certain degree of mathematical ability is essential for maximisation and since the Warli (a tribal community of Thane district in Maharastra do not possess such a required level of mathematical knowledge, they will not be able to maximise their gain and hence their behaviour cannot be said to be rational. Apparently Deshpande's rationality concept is one of maximisation according to economic theory norms.

### 4.4.2 Production Conditions

Let us now summarise the production conditions under which the Chotanagpur farmers take decisions which we have examined in great detail in various sections of Chapter 2 and Chapter 3.

(1) The farmers are producers as well as consumers of the crop. Farmers' decision as a producer is sometimes affected by his own decision as a consumer. Tribal farmers generally grow ragi for their subsistence irrespective of prices and yield rates. The poor farmers may not also spare their land for non-subsistence crops despite relatively higher returns for these crops because they first want to make sure of getting minimum for their subsistence.

(2) The farmers produce primarily food crops mainly cereals using traditional agricultural practices. The farmers (both tribal and non-tribal) are, however, using the new agricultural practices in growing wheat in recent years. Agriculture is the main source of their livelihood. A fraction of their output is, however, sold in the market for purchasing other non-agricultural commodities required for them.

(3) All operational holdings are owned and self-operated in all the districts except in Hazaribagh where only 1.31 per cent area is partly owned and partly rented as per the 1971 Agricultural Census.

(4) Kharif crops are generally grown under rainfed conditions whereas rabi crops specially HYV wheat requires regulated water supply through secured irrigation facilities. Moreover, traditional variety wheat, barley and gram can also be grown and are usually grown under rainfed conditions.

(5) Agro-economic conditions required for growing winter rice precludes the possibilities of growing other crops on the same land during the same kharif season. Since winter rice is the major crop grown during the kharif season the substitution possibility is very much limited. However, there is no such limitation in growing the crops during the rabi season.

(6) The area that is suitable for growing winter rice is very much limited. The conversion of upland into low land for growing winter rice is a very costly affair and not possible for the poor farmers. The area under other crops can be increased through substitution and also by utilising the fallow lands because a considerable proportion of total cultivable area remains uncultivated every year (about 43.59 per cent during 1975-76) in Chotanagpur region either due to the want of sufficient finance or due to the want of sufficient rainfall.

(7) The practice of crop-rotation system followed by the Chotanagpur farmers is of course a notable feature of their agriculture. It can also be looked upon as a systematic crop substitution.

Such are the conditions under which the crops are grown in Chotanagpur. It remains to examine these conditions with the aid of economic analysis, so as to isolate the factors determining farmers' decisions and to discover the direction of influence of these factors under changing conditions viz. before and after the introduction of HYV technology in agriculture.

## 4.5 HYPOTHESES BEHIND THE SIMPLE MODEL USED IN THIS STUDY

Our model is based on the standard hypothesis of "simple working rule" reflecting informal decision making process noted earlier namely the amount of land given to an individual crop(s) in one period is determined on the basis of the area allotted to the crop(s) in the previous period along with adjustments in response to changes in prices and other factors. In other words, the traditional farmers take into account the area allotted to different crop(s) for the previous year while deciding on the allocation of land to different crop(s) in the current year and only adjustments are made in response to changes in prices, weather conditions and irrigation facilities etc.

The hypothesis of simple working rule leads to the heuristic formulation that acreage planted under the crop in the current period is a function of area planted in the previous period and the price of the crop in previous period. Assuming a log-linear relationship between the variables we can specify this as

$$\text{Log } A_t = a + b \text{ Log } A_{t-1} + c \text{ Log } P_{t-1} + u_t$$

where,

$A_t$ is area planted in year t,

$P_{t-1}$ is price in year t-1, and

$u_t$ is the usual error term.

It should be noted that the above equation is schematically the same as that of the reduced form equation of the Nerlovian partial adjustment model (see equation 5 in - Appendix-IV), the only difference lies in the interpretation of the generation-process of the lagged dependent variable and its coefficient. In the Nerlovian framework it is the outcome of the adjustment mechanism assumed in the model whereas we have taken it as a traditional variable. Again there is no reasonable interpretation of the negative coefficient of the lagged dependent variable in the Nerlovian model (because of the assumption $O<b_4\leq 1$ which implies positive coefficient of the lagged dependent variable) but it has its own meaning in our model and (as we shall see later) it reflects some traditional characteristics of the farmers.

One can also find a similar rationale behind the inclusion of lagged dependent variable in demand analysis. Houthakker and Taylor [58] argue that the quantity demanded in any one period depends, among other things on the quantity demanded in previous periods due to a habit formation process which is a characteristic of human behaviour. They termed this behavioural hypothesis "the habit formation principle". A traditional farmer usually goes on producing certain crops every year for their subsistence and only marginal changes in area are expected in response to changes in prices and other related factors.

In the next section we try to develop an analytical model which will help us in identifying suitable shifter variables, in traditional agriculture as well as in modern agriculture which might affect the acreage-price association.

## 4.6 ANALYTICAL MODEL

Assume that the farmer is initially in a position of equilibrium with a given allocation pattern that might have been built up over the years. What are the factors that might cause him to change the existing allocation pattern and hence to seek a new position with better allocation of land ? The answer to this question may enable us to identify the factors that determine the farmers' decision in respect of allocating the land to different crops.

Many different types of incentives might cause the farmer to change his existing allocation pattern. Among the various incentives, increase in the prices of output may be the foremost to the extent that the production is market-oriented. The incentives may also be in the form of secured irrigation facilities and in the wake of the introduction of HYV technology the incentives may be more importantly in the form of the availability of the new inputs like HYV seeds, chemical fertilisers, insecticides etc., and, of course, the technical know-how for proper use of the new inputs. It should be noted, however, that the farmers' response to price changes may be subject to the availability of suitable weather conditions in general and the availability of the basic infrastructural facilities needed for adopting the new agricultural practices in the HYV period.

Let us proceed with the assumption that (i) the farmer is producer as well as consumer of the product, (ii) the farmer takes decisions under uncertainties, (iii) the substitution possibility is strictly limited which implies that the increase in area under the crop is possible only through (a) utilisation of fallow land which remain uncultivated due to want of money or good weather conditions, (b) reclamation of culturable waste land other than fallow land and (c) reclamation of deforested land, (iv) the farmer uses traditional agricultural practices and the technology remains unchanged over the period.

A simple model may be introduced, which, within the framework of these simplified assumptions, will be helpful in the effort to discover the factor influencing farmers' decision in allocating the land to different crops under traditional agriculture. By relaxing the last two assumptions; specially the last one, we can allow the model to be more realistic and consistent with the changed circumstances and thus it will be possible for us to determine the factors influencing the farmers' decision-making under the influence of the impinging modern agriculture.

Under the simplified assumptions we can proceed with the arguments that a farmer has planted a certain crop under rainfed conditions in a fixed amount (say normal) of land in year t. Assume that the farmer expects the price to rise, he will try to increase the area under the crop in the next year (i.e. year t + 1) upto the extent he could manage the additional land under good weather conditions and in bad weather conditions it is very probable, that he may not plant the crop or only a fraction of land could be planted due to the want of sufficient rainfall during sowing season. But in the next year (i.e. year t + 2) the farmer will try to plant the crop at least on the land in which he has planted in the year before the last (i.e. in year t) under good weather conditions if the crop is of subsistence character. What we want to emphasize through these arguments is that the farmer will try to maintain the area under the crop normally planted if the crop is grown for his subsistence.

Under these conditions we find a great scope for the price incentive to be effective and weather factors may be more important in farmers' decision-making because even if the farmer wants to increase the area induced by the price rises he may not be in a position to increase the area under the crop if weather becomes unfavourable and under good weather conditions the farmer may also increase the area even if he expects the price to decline, provided he is able to manage some additional land; otherwise he will try to maintain at least the area which he normally planted. The farmers' behaviour appears to be non-normal but quite consistent under the framework of the analysis.

It is quite possible that the tribal farmer because he wants to meet his current requirements only and not accumulate and also because of his limited taste and natural conservatism, might respond to prices of the output and more importantly to the relative price of his output i.e. terms of trade, but in a perverse way. If the price of his output relative to other commodities (mainly non-agricultural) falls he will require additional output even to maintain his present level of consumption of other goods and thus he might produce more. If the price of his output relative to others increase he will not require additional output. Since his cash requirements are primarily fixed he might reduce his output and prefer leisure. All this indicates the backward sloping producers' response to price changes, rather than the normal forward sloping supply response appearing in the usual producer behaviour analysis.

Such supply response is perverse in the sense that it is counter intuitive in the profit maximisation sense. Intuitively one would expect a rise in the prices of outputs relative to those of goods purchased to result in producers expanding both output and quantity sold. If they were to do exactly the opposite when prices move in their favour, their behaviour could be called perverse.

Such perverse producers' response to price changes for tribal farmers, though of specific applicability, may neither be unrealistic nor irrational as argued earlier. Such backward sloping supply curves in a backward setup and even their formal mathematical specifications are not unknown in economic literature. Such responses have also been noted in the context of marketed surplus by Khatkhate as the following passage indicates. "Suppose the prices of agricultural output rise in relation to non-agricultural prices. Since money requirements are a priorily fixed charge on farmers' output, he would be prone to market a smaller proportion of his output for cash in view of the price rise, and consequently the consumption of his output, which is already sub-normal, would increase. But the rise in prices would not have any impact on production as he is already producing to the hilt to raise his consumption level .... The position is reversed when prices of agricultural goods fall " [69 p.189]. Dean has also noted that "Relatively high tobacco prices are quite likely to lead to high levels of tobacco production. But since the income effect works in the opposite direction, it cannot be claimed that a contrary result would be inconsistent with economic theory" [33 p.45].

Now let us relax assumption (iii) and allow the substitution possibilities to operate into the model. In this case the change in area under an individual crop is quite possible irrespective of the physical limitation of the total area at the farmers' disposal. A rational farmer is expected to substitute the crop for which he expects a higher price relative to other substitutable crop(s). A farmer might also respond to total return from the crop rather than to price alone. He might also consider the relative risk in return (risk is measured by the coefficient of variation) of the crop. If the crop is grown solely for his own consumption then tradition and quality consideration may be more important than the price consideration for him. Besides these considerations the farmers' decision of substituting one crop for another may also be influenced by some technical factors such as water requirements of the crops in relation to its availability and usage of the crop-rotation system.

Autumn rice grown during kharif season requires more water than maize or ragi, therefore a rational farmer is expected to plant autumn rice under good weather conditions and maize or ragi under scanty rainfall condition. Weather conditions may also sometimes compel the farmer to substitute one crop for another without any prices considerations coming in to the picture and negative price response may be the result. Though this appears rather peculiar it is consistent in the present framework of our analysis. Similarly wheat requires secured irrigation facilities therefore if irrigation facilities are made available in sufficient amount the farmer might prefer to grow wheat and substitution for wheat as against gram and barley grown during rabi season may also be there. Substitution decisions of the farmer may also be influenced by the crop-rotation system followed by the Chotanagpur farmers as a form of systematic crop substitution. It should be noted that the short-run changes in relative prices in small magnitude may not be sufficient for inducing him to change his tradition-dictated crop-rotation practice. If this is so then one should not expect positive price response from the traditional farmers to the extent crop-substitution prevails unless price changes involved are very large and persisting ones.

Let us now allow the technology to be changing over the period and proceed with the assumption that (a) the adoption of new agricultural practices necessarily leads to a higher level of production, (b) constant returns to scale operate initially and (c) the new technology is size-neutral and the increased return from growing the new variety exceed the cost.

The farmer is now in a position to use the HYV seeds. It is quite possible that the farmer will replace traditional variety seeds by their new variety counterparts in the existing cultivated land or he may prefer to grow it in additional land which may come from substituting other competing crops for which HYV seeds are not available or it might come from utilising fallow lands because less fertile land can now be used for growing HYV seeds with appropriate dose of chemical fertilisers and insecticides under irrigated conditions. These different alternatives would lead to different absolute and relative area allocations. It should also be noted, however, that the use of new agricultural practices require sufficient capital for purchasing new inputs like HYV seeds, chemical fertilisers etc. HYV wheat in particular requires regulated water supply through secured irrigation facilities. However, the use of HYV seeds of wheat under rainfed conditions is not ruled out though it becomes very risky in the sense that HYV seeds are very much sensitive to chemical fertilisers [133] and the use of fertilisers requires sufficient water. If rainfall does not occur in time and in sufficient amount, the yield may be affected severely and sometimes the whole crop may be destroyed.

Since basic infrastructural facilities needed for the adoption of new agricultural practices are lacking to a great extent in the Chotanagpur region, one should not expect very high price response even in the HYV period. In fact, in the absence of such infrastructural facilities the farmer may not also respond to price changes despite his desire. The government is now very much seized about these difficulties and is trying to provide the infrastructural facilities gradually. At the initial stages when the farmer gets these facilities he might be interested in growing new variety seeds without any price consideration i.e. he might solely respond to these facilities specially in the case of those crops for which the HYV seeds are available. He may also be influenced by price incentives but the other facilities in the form of availability of necessary inputs might dominate his decision in the sense that his decision may not be implementable unless the necessary facilities are made available to him. In the long-run when these facilities become available in sufficient amount, which would also by themselves, because of the external contact they bring with them, or along with other modernising influences lead to a modern-oriented change in the farmers' attitude and norms, the farmer may readily respond to price changes relative to the prices of inputs. A farmer is expected to grow the crop or substitute it for other competing crops only upto the extent he expects positive net return and relatively higher net return respectively (the assumption that the increased return from growing the new variety would exceed the cost may not hold true in the long-run). As commercialisation increases and assurance of provision of foodgrains of consumption when needed is there, the farmer would make less and less distinction between food and subsistence crops and others from the point of view of supply response.

The above discussion sufficiently indicates that a number of factors influence the farmers' decision about allocating their land to different crops. In the light of the above discussion we specify these factors in the form of suitable variables for incorporating them in our regression analysis. It should be noted that the above model is developed on the individual-farmer basis but it can also be used for a group of farmers, say the farmers in the district, on the assumption of homogeneity of the decision making units.

## 4.7 SELECTION OF VARIABLES

Out of a number of factors influencing farmers' acreage allocation decision suggested in the model, we have selected for our multiple regression analysis only those which could be fairly important and for which time-series data are available. However, other factors are also considered at appropriate places in subsequent chapters while dealing with different aspects of acreage responses. First we shall discuss the criterion used for selecting the dependent variable and then consider the selection of independent variables.

### 4.7.1 **Dependent Variable**

Supply response to price changes is indirectly examined through the acreage response to prices. Planted area, as dependent variable is generally taken as a better decision variable by most researchers because area under the crop is in most cases, under the direct control of the farmers but this is not the case with the actual production which is governed by the application of inputs and the biological processes involved; unlike the factory processes the latter are not fully under control or fully predictable: Backward agriculture, in particular, has to face a great deal of risk. Even the agricultural research stations equipped with all modern facilities and efficient management fail to control the effects of factors like ground frost and hail storms.

In fact planned output may be defined as the product of planned area and planned yield rate, in which case the elasticity of planned output is equal to the elasticity of planned acreage plus the elasticity of planned yield rate. An approximation of planned output by planned area thus indirectly may assume that (a) inputs other than land are varied in proportion to acreage and returns to scale are constant, and (b) the correlation coefficient between planned output and planned area is very near one, which implies that the elasticity of planned yield rate to price in question is almost zero, which may mean that the yield rate remained constant in the relevant period.

Defending the first assumption Raj Krishna [71] argues that this may be a reasonable assumption for poor regions, where the rural manpower finds gainfull employment for a part of the year, and, therefore, labour-input can be increased with the land input without much additional paid out cost, and, at the prevailing level of technique, the cost of capital service per unit of land is also small. This argument seems valid for Chotanagpur. The yield rate of most of the crop is mainly influenced by the weather factors in Chotanagpur as indicated by the very small and insignificant productivity-trend coefficients noted earlier in Chapter 3. Even in the case of the wheat crop for which the use of new agricultural practices is indicated we find that the rate of growth of area under the crop is much higher than the rate of growth of productivity indicating Chotanagpur farmers' preference for extensive farming techniques rather than intensive farming techniques. Thus area can reasonably be taken as a proxy to output and acreage response may be considered as a rough measure of the supply response of Chotanagpur farmers.

### 4.7.2 Independent Variables

In our multiple regression model explaining variation is area under the crop(s), we have considered the preceeding years' area, preceeding years' price, preceeding years' yield-rate, preceeding years' gross return per unit of area, current years' pre-sowing rainfall, irrigated area and fertiliser prices as other explanatory variables.

Since the farmers sell a bulk of their marketable surplus just after the harvest, farm harvest price seems to be the most reasonable approximation to the price actually realised by farmers. Therefore, absolute farm harvest price with one period lag is taken as one of the independent variables. Generally farmers take their decisions on the basis of previous years' price because they expect that the previous years' price would continue at least in the coming season. This is similar to the Nerlovian naive price expectation. While substituting one crop for another the farmers might use the relative price rather than the absolute price of the crop; therefore own price relative to the price of its substitute crop is taken as an independent variable while examining the crop substitution behaviour. Since the farmers purchase many goods specially non-agricultural goods after selling their products, the All Commodity Wholesale Price Index has been used as deflator for computing relative agricultural prices with weighted average price of total cereals being used as the representative of agricultural prices. We are aware of the fact that such a deflator may over estimate the farmers' demand of goods and thus it may underestimate the price which farmers actually receive. But in the absence of any consumer price index of subsistence farmers of this region, it was thought that such a deflated price would be closer to real prices received by subsistence farmers and therefore used despite its inadequacies. Some other researchers have also used such an index as a deflector in their respective studies. Two such studies are Oury [99] and Sawant [117].

If the farmers produce the crops mainly for the market, price may be an important consideration for them but the subsistence farmers may attach more importance to the yield rate because most of the crops are grown for their own consumption. Again the price may be the only consideration if yield rate remains the same over the period. If the yield rate varies then price may not be the sole decision factor. This would be so specially in the HYV period, when technological development takes place, the year-to-year changes in the land allocation reflect therefore changes in farmers' decisions not only due to variation in prices but due to variation in the yield rate as well. Hence it is necessary to consider the yield variable also while studying farm supply responses. Previous years' yield rate of the crop(s) is considered as another important variable in the regression used here.

A producer-consumer farmer might also consider price and yield rate both simultaneously and may be interested in income from the crop specially net cash income. In the absence of cropwise cost of production information, the gross return per hectare is considered in this study as one of the independent variables and included in the regression as an alternative to the price and the yield rate variable because gross return is obtained by multiplying the yield rate of the crop with their respective farm harvest prices. The importance of gross return as an independent variable is also realised by Sawhney [118] and Narayan and Parikh [89] and used in their respective studies.

Weather is another important variable considered in our acreage response equations. Almost all important cereals are grown under rainfed conditions in the Chotanagpur region and here lies the importance of the weather factor in determining farmers' decisions in respect of acreage allocation. Pre-sowing rainfall (absolute amount in millimetres) is taken as a proxy for the weather factor. This method is also used by some other Indian writers. The particular months included are guided by a priori knowledge of the crop growth characteristics and of the general climatic conditions in the region. Generally the kharif crops are sown just after the arrival of monsoon. Therefore, the rainfall during the period from May to July is included in pre-sowing rainfall for the crops like autumn rice, maize and ragi. Generally the broadcast method of sowing winter rice is completed by July but transplantation work may continue upto the last week of August; therefore, rainfall for the months of June, July and August is taken as pre-sowing rainfall for winter rice. For rabi crops like wheat, barley and gram the rainfall for the period from September to November is included. In fact rainfall is but only one component of weather conditions and it would have been better if we could prepare the composite weather index for this purpose as suggested by Sidhu and Kaul [127] but it could not be done for such a variety of crops and for different districts under consideration.

Total irrigation facilities made available to the farmers during abi season are introduced as a shifter variable in the functions explaining acreage because changes in the total irrigation facilities available might easily dictate changes in the acreage of irrigated crops. Total irrigated area under important rabi crops (like wheat, barley and gram) are taken as a proxy to the irrigation facilities made available to the farmers in the season assuming that all the facilities available in the area are fully utilised by the farmers. Since irrigation facilities are still at a development stage and farmers expect more irrigation facilities than what they had in the previous year and take their decision accordingly, the irrigation facilities in the current year are to be preferred as a shifter variable to the irrigation facilities made available to the farmers in the previous year.

We expect that input prices, in general, do not affect acreage decision of the farmers of Chotanagpur because they mostly use non-marketed inputs except for purchasing HYV seeds and fertiliser etc., in the HYV period for using them in growing wheat. In the absence of cost of production data we have used the index of fertiliser price, for simplicity, as one of the explanatory variables in wheat acreage response equation in the HYV period only. The price of urea (the commonly used fertiliser in the region) at all India level, which was easily available, is used for this purpose in the absence of such information at state level or at district level.

It may be noted that no adjustment-mechanism is associated with the model to explain previous years' area (lagged dependent) as one of the explanatory variables. Rather it is based on the hypothesis of simple working rule and informal decision making process as discussed above. It also takes into account to some extent the trends, if any, in the variables included in the regression.

## 4.8 ESTIMATION METHOD AND STATISTICAL TEST PROCEDURES USED IN THIS STUDY

Single equation model with Ordinary Least Squares (OLS) estimates is used for estimating the parameters in the proposed response models. The log-linear form equation is preferred to the simple linear form because of its implications of multiplicative effects; besides the coefficients give a direct measure of elasticities. The relative importance of explanatory variables in explaining the variations in dependent variable is determined by the magnitude of the relevant coefficients. Two-tailed t-test has been used for testing the significance of these coefficients. Error terms are assumed to be normally, identically and independently distributed with zero mean. These assumptions regarding the error terms hold in all the regressions estimated and presented in this thesis $R^2$ the square of multiple correlation coefficient and F-values for testing the significance of the explained sum of squares are computed and presented with the regression result. It should be noted that t and F tests have only asymptotic validity because of the presence of the lagged endogenous variable in the equation. The standard Chow test has been used for testing the overall homogeneity of regressions between the two sub-periods and also between the districts. This test is also affected by the presence of lagged endogenous variable in the equation.

It should be noted that the OLS estimation procedure is not the best in the small sample case when lagged dependent variable enter the regression equation as independent variables. In small samples the least squares estimates are biased, although they are consistent when the errors are serially independent. Because of the bias in the estimates the computed standard errors are also biased. "These problems associated with least squares estimation do not mean, however, that it is an inappropriate procedure. No other estimation procedure has been shown to be better in small samples when lagged dependent variables are on the right hand side of the equation" [108 p.176].

We have employed the Durbin-Watson test [47] for testing the presence of serial correlation in the error terms. In fact Durbin-Watson statistics should not be used as an exact text in this case due to the presence of lagged dependent variable as explanatory variable but in the absence of any other exact test, we have used it despite its limitations, as an approximate test for the purpose. Moreover, Durbin [46] has modified the test for using it when lagged dependent variable is present among the explanatory variables but unfortunately it is a large sample test and we have only 20 observations at the most. It should be noted that the Durbin-Watson test is used only for the combined period data with 20 observations and not for the two sub-periods each of which contain only 10 years' data. The null hypothesis of no serial correlation is accepted at 5 per cent level of significance in all the cases except in a few cases (like ragi and wheat for Palamau and total cereals for Hazaribagh) for which the test becomes inconclusive.

## 4.9 ALTERNATIVE SPECIFICATIONS OF THE REGRESSION EQUATIONS USED

Alternative specifications were needed for studying the various aspects of supply responses in Chotanagpur region. They were also needed because of the differences in the agro-economic characteristics of the crops grown and because of the technological changes over the period. The alternative specifications of the multivariate regression equations used in our analyses along with the explanation of the symbols used in the equations are given below.

**A: Acreage Response Equations**

$$A.1 \text{ Log } A_{rt} = a + b_1 \text{ Log } A_{rt-1} + b_2 \text{ Log } P_{rt-1} + b_3 \text{ Log } Y_{rt-1} + \text{ Log } R_{rt} + u_t.$$

$$A.2 \text{ Log } A_{rt} = a + b_1 \text{ Log } A_{rt-1} + b_2 \text{ Log } G_{rt-1} + b_3 \text{ Log } R_{rt} + u_t$$

$$A.3 \text{ Log } A_{rt} = a + b_1 \text{ Log } A_{rt-1} + b_2 \text{ Log } G_{rt-1} + b_3 \text{ Log } R_{rt} + b_4 \text{ Log } F_t + u_t$$

$$A.4 \text{ Log } A_{rt} = a + b_1 \text{ Log } R_{rt} + b_2 \text{ Log } I_t + u_t$$

$$A.5 \text{ Log } \frac{A_{rt}}{A_{rt-1}} = a + b_1 \left( \text{Log } \frac{P_{rt-1}}{P_{rt-2}} \right) Z_1 + b_2 \left( \text{Log } \frac{P_{rt-1}}{P_{rt-2}} \right) Z_2 + b_3 \text{ Log } \frac{R_{rt}}{R_{rt-1}} + b_4 \text{ Log } \frac{I_t}{I_{t-1}} + b_5 \text{ Log } \frac{A_{rt-1}}{A_{rt-2}} + u_t$$

or Simply

$$a_{rt} = a + b_1 e_1 + b_2 e_2 + b_3 r_{rt} + b_4 i_t + b_5 a_{rt-1} + u_t$$

$$A.6\ \text{Log}\frac{A_{rt}}{A_{rt-1}} = a + b_1\left(\text{Log}\frac{G_{rt-1}}{G_{rt-2}}\right)Z_1$$

$$+ b_2\left(\text{Log}\frac{G_{rt-1}}{G_{rt-2}}\right)Z'_2 + b_3\,\text{Log}\frac{R_{rt}}{R_{rt-1}}$$

$$+ b_4\,\text{Log}\frac{I_t}{I_{t-1}} + b_5\,\text{Log}\frac{A_{rt-1}}{A_{rt-2}} + u_t$$

or Simply

$$a_{rt} = a + b_1\acute{e}_1 + b_2\acute{e}_2 + b_3 r_{rt} + b_4 i_t + b_5 a_{rt-1} + u_t$$

**B: Overall Acreage Response Equations**

$$B.1\ \text{Log}\,A_t = a + b_1\text{Log}\,A_{t-1} + b_2\text{Log}\,P_{t-1} + b_3\text{Log}\,Y_{t-1} + b_4\text{Log}\,R_t + u_t$$

$$B.2\ \text{Log}\,A_t = a + b_1\text{Log}\,A_{t-1} + b_2\text{Log}\,P^R_{t-1} + b_3\text{Log}\,Y_{t-1} + b_4\text{Log}\,R_t + u_t$$

$$B.3\ \text{Log}\,A_t = a + b_1\text{Log}\,A_{t-1} + b_2\text{Log}\,G_{t-1} + b_3\text{Log}\,R_t + u_t$$

**C: Cropping Pattern Response Equations**

$$C.1\ \text{Log}\left(\frac{A_{rt}}{A_{st}}\times 100\right) = a + b_1\,\text{Log}\left(\frac{A_{rt-1}}{A_{st-1}}\times 100\right)$$

$$+ b_2\,\text{Log}\left(\frac{P_{rt-1}}{P_{st-1}}\times 100\right)$$

$$+ b_3\,\text{Log}\left(\frac{Y_{rt-1}}{Y_{st-1}}\times 100\right) + u_t$$

$$C.2\ \text{Log}\left(\frac{A_{rt}}{A_{st}}\times 100\right) = a + b_1\,\text{Log}\left(\frac{A_{rt-1}}{A_{st-1}}\times 100\right)$$

$$+ b_2\,\text{Log}\left(\frac{G_{rt-1}}{G_{st-1}}\times 100\right) + u_t$$

**D: Overall Cropping Pattern Response Equations**

D.1 $CV(A_t) = a + b\,CV(P_{t-1}) + u_t$

D.2 $CV(A_t) = a + b_1\,CV(P_{t-1}) + b_2 I_t + b_3\,CV(A_{t-1}) + u_t$

In the above equations

$A_{rt}$ = Area under rth corp (in hectare) in year t.
(r = 1, 2,....m, t = 1, 2,.......n)

$A_t = \sum_{r=1}^{m} A_{rt}$ = Aggregate area under total cereals in year t.

$A_{st}$ = Area under the substitute (of rth crop) crop in year t.

$P_{rt-1}$ = Farm harvest price (in Rs. Per Qtl.) of rth crop in year t-1

$$P_{t-1} = \sum_{r=1}^{m} P_{rt-1} \times O_{rt-1} \Big/ \sum_{r=1}^{m} O_{rt-1}$$

= Weighted average price of total cereals in year t-1 weights being the Output (O) of the crops.

$P^R_{t-1}$ = Relative price of total cereals in year t-1 i.e. weighted average price of total cereals relative to the Economic Advisors All Commodity Wholesale Price Index.

$Y_{rt-1}$ = Yield rate of r th crop (in Kg. per Ha.) in year t-1.

$$Y_{t-1} = \sum_{r=1}^{m} O_{rt-1} \Big/ \sum_{r=1}^{m} A_{rt-1}$$

= Average yield rate of total cereals in year t-1.

$Y_{st-1}$ = Yield rate (in Kg. per Ha.) of substitute (of rth crop) crop in year t-1.

$G_{rt-1} = P_{rt-1}\, Y_{rt-1}$ = Gross return (in Rs. per Ha.) of r th crop in year t-1.

$G_{t-1} = \sum_{r=1}^{m} G_{rt-1}$ = Average gross return of total cereals in year t-1.

$G_{st-1}$ = Gross return of substitute crop in year t-1.

$R_{rt}$ = Pre-sowing rainfall (in millimetres) relevant to the r th crop in year t.

$R_{t}$ = Pre-sowing rainfall (in millimetres) relevant to total cereals.

$I_{t}$ = Total irrigated area under important crops during rabi season in year t.

$F_{t}$ = Index of fertiliser price (1967-68 = 100) in year t.

$$CV(A_t) = 100\left[\frac{1}{m}\sum_{r=1}^{m}\left(A_{rt} - \overline{A}_t\right)^2\right]^{1/2} \Big/ \overline{A}_t$$

= Coefficient of variation (in percentage) of area under rth crops in year t i.e. Index of the overall variation in the cropping pattern.

$$CV(P_{t-1}) = 100\left[\frac{1}{m}\sum_{r=1}^{m}\left(P_{rt-1} - \overline{P}_{t-1}\right)^2\right]^{1/2} \Big/ \overline{P}_{t-1}$$

= Coefficient of variation (in percentage) of farm harvest prices of r th crops in year t-1, i.e. Index of the overall variation in the price pattern.

Z = Dummy Variables :

$$Z_1 = 1 \text{ if } \frac{P_{rt-1}}{P_{rt-2}} > 1\,;\ Z_2 = 1 \text{ if } \frac{P_{rt-1}}{P_{rt-2}} \leq 1$$

= 0 if otherwise = 0 if otherwise

and

$$Z'_1 = 1 \text{ if } \frac{G_{rt-1}}{G_{rt-2}} > 1\,;\ Z'_2 = 1 \text{ if } \frac{G_{rt-1}}{G_{rt-2}} \leq 1$$

= 0 if otherwise = 0 if otherwise

$u_t$ = Random error term (t = 1, 2.... n), normallly identically and independently distributed with zero mean.

Equation A.1 to A.4 are used for estimating the acreage response of individual crops (equation A.3 and A.4 are used only for wheat because of its specific characteristics distinguishing it from other cereals specially in the HYV period) whereas overall acreage response of total cereals is examined in terms of equation B.1 to B.3. Non-symmetric acreage response to price and gross returns is alternatively examined in the case of two major cereals (winter rice and wheat) using equation A.5 and A.6. Equation C.1 and C.2 are used for studying the crop substitution behaviour taking the pairs of substitute crops whereas equation D.1 and D.2 are employed for studying the overall cropping pattern response and factors underlying year-to-year variation in the overall cropping pattern in the districts of Chotanagpur region.

## APPENDIX IV

### A BRIEF NOTE ON NERLOVIAN SUPPLY RESPONSE MODEL

Marc Nerlove [92] has attempted to estimate the elasticity of supply for corn, cotton and wheat in the USA. His work has dealt primarily with the role which the farmers' expectations of future prices play in shaping their acreage allocation (to different crops) decisions. He has used the concept of distributed lag in formulating the expectation models in his attempt to arrive at certainty equivalents for uncertain future. The essential ideas of the Nerlovian

Model are that

(1) Farmers keep adjusting their output[2] over time towards a long-run desired level of output, on the basis of expected future prices.

(2) Expected future price is some notional value which can be approximated by a weighted moving average of the past prices with the weights declining with the time-lag and

(3) Short-run adjustments in the output, which are made keeping the long-run desired level of output in mind, may not fully reach the long-run desired level because of various constraints on the speed of adjustment.

Nerlove's model basically consists of three equations

(1) $A^*_t = a + b_1P^*_t + b_2Z_t + u_t$

(2) $P^*_t - P^*_{t-1} = b_3(P_{t-1} - P^*_{t-1})$

or $P^*_t = b_3P_{t-1} + (1-b_3)P^*_{t-1}$ $\qquad 0 < b_3 \leq 1$

(3) $A_t - A_{t-1} = b_4(A^*_t - A_{t-1})$

or $A_t = (1-b_4)A_{t-1} + b_4A^*_t$ $\qquad 0 < b_4 \leq 1$

Where,

$A^*_t$ = Long-run desired acreage of the crop at time t.

$A_t$ = Actual acreage at time t.

$P_t$ = Actual price at time t.

$P^*_t$ = Expected normal price at time t.

---

2. Area is a proxy for output in the Nerlovian model.

$Z_t$ = Other exogenous factor (s) affecting supply at time t.

$u_t$ = A random residual.

$b_3$ = Price expectation coefficient.

$b_4$ = Acreage adjustment coefficient.

Equation (2) presents the process of adaptive price expectation and indicates that the current expected price $P^*_t$ falls somewhere in between the last years' actual price ($P_{t-1}$) and the last years' expected price ($P^*_{t-1}$) depending upon the elasticity of expectation. If $b_3$ is zero, the actual prices are totally divorced from expectation implying that there exists only one expected price for all time periods, while a unitary value implies a naive cobweb-type model where expected prices are identical with previous years' actual price.

Acreage adjustment mechanism is presented in equation (3). The coefficient $b_4$ indicates how fast the farmers are adjusting to their expectations. A value of $b_4$ close to zero would mean that the farmers are very slowly adjusting to the changing prices whereas the value of $b_4$ close to 1 would mean that the farmers are quickly adjusting to the price changes. A meaningful interpretation thus requires that $0 < b_4 < 1$.

Allowing both adaptive price expectation (equation 2) and area adjustment (equation 3) to operate into the model, the reduced form equation thus obtained is called the Nerlovian full adjustment model i.e.

$$(4)\ A_t = ab_3b_4 + b_1b_3b_4 P_{t-1} + b_2b_4 Z_t - b_2b_4 (1-b_3) Z_{t-1}$$

$$+ [ (1-b_3)+(1-b_4) ] A_{t-1} - (1-b_3)(1-b_4) A_{t-2}$$

$$+ b_4u_t - b_4 (1-b_3) u_{t-1}$$

If $b_3$ is assumed to be unity, the reduced form equation becomes simpler and known as the Nerlovian partial adjustment model i.e.

$$(5)\ A_t = a\, b_4 + b_1\, b_4\, P_{t-1} + (1-b_4) A_{t-1} + b_2\, b_4\, Z_t + b_4\, u_t$$

Various estimation problems arise while estimating the Nerlovian full adjustment model. One such problem arises because of the presence of auto-correlation in the error term which is clearly evident in the reduced form equation itself (see equation 4). This brings in as is well known in-consistency in the CLS estimates. Some of these estimation difficulties have, however, been resolved by later writers such as Behrman [13] and Nowshirvani [95] who used maximum likelihood estimating techniques and Askari and Cummings [5] who impose certain restrictions on one or other structural coefficient. Besides this, many other problems are encountered when a model of Nerlovian type is used for annual crops and it has not worked well for perennial crops specially livestocks [6]. However, Nerlove, Grether and Carvelho [94] have recently developed a new model for United States cattle industries based on the dynamic op-timisation principle.

FIVE

# ACREAGE RESPONSE IN CHOTANAGPUR REGION
## Inter District and Inter Temporal Variations

### 5.1 INTRODUCTION

In this chapter we examine the supply response of individual crops (cereals) in terms of acreage elasticity to price and non-price factors. The analysis of supply response for individual crops is of special importance if the crop is grown in isolation to the other crops i.e. if the substitution possibilities are limited due to physical or technical constraints. Again if some special crop-wise drive is made as in the case of wheat in the HYV period, the crop-wise analysis would have some additional advantage so far as the problem and prospect of increasing the output of specific crop is concerned. It would be interesting to know, in general, what impact, if any, price and non-price factors have on acreage response of individual crops in the two periods.

Non-symmetric acreage response to price changes for subsistence crops is hypothesized and examined in the case of two major cereals viz. winter rice, the major kharif cereal, and wheat, the major rabi cereal grown in the Chotanagpur region. Apart from the acreage response of individual crops we have also discussed the overall acreage response to overall prices and other non-price factors in an attempt to examine the nature and extent of agricultural supply response to agricultural prices in a backward economy like Chotanagpur region.

## 5.2 ACREAGE RESPONSE OF INDIVIDUAL CROPS : SIX IMPORTANT CEREALS

### 5.2.1 Introduction

We have examined the acreage response of all important cereals grown in the Chotanagpur region. The crops included in this study are winter rice, autumn rice, maize and ragi grown during kharif season and wheat and barley grown during the rabi season. Winter rice is the major kharif cereal whereas autumn rice, maize and ragi are minor cereals as compared to winter rice. Wheat is the major cereal grown during the rabi season. In the next section we discuss the alternative specifications of the regression equations determining the relationship between area planted under the individual crops and various factors influencing the decision of the traditional farmers. We are using the working rule hypothesis that farmers base their decisions on what they did in the previous year and make adjustments on the basis of the levels current or expected as the case may be of the relevant variables, discussed earlier.

### 5.2.2 Alternative Specifications of the Regression Equations

Following the simple working rule hypothesis one would expect that the traditional farmers, while deciding on the area to be allotted to a specific crop, might consider the area allotted to the specific crop in the previous year and try to adjust the area in the current year in response to the changes in prices and some other relevant factors which might influence their decisions. Therefore, in our regression equations we have included previous years' area, previous years' price, previous years' yield rate and previous years' gross return and current years' pre-sowing rainfall, fertiliser prices and total irrigated area during the rabi season as independent variables. On the basis of the earlier discussion we find these variables to be relevant.[1] We have considered log-linear relationships with some

---

1. Since these are regional data, the population variable could also be relevant. Though we have not included it as a separate variable as for each of the two periods only one independent population estimate (the census estimate) is available, to some extent the lagged endogenous variable takes care of this as population in this region is a monotonically increasing variable leading to a possible increasing trend in the endogenous variable.

combinations of these variables as independent variables. The following are the alternative equations which have been estimated.

$$\text{A.1 } \text{Log } A_{rt} = a + b_1 \text{Log } A_{rt-1} + b_2 \text{ Log } P_{rt-1} + b_3 \text{Log } Y_{rt-1} + b_4 \text{Log } R_{rt} + u_t$$

$$\text{A.2 } \text{Log } A_{rt} = a + b_1 \text{Log } A_{rt-1} + b_2 \text{ Log } G_{rt-1} + b_3 \text{Log } R_{rt} + u_t$$

$$\text{A.3 } \text{Log } A_{rt} = a + b_1 \text{Log } A_{rt-1} + b_2 \text{ Log } G_{rt-1} + b_3 \text{Log } R_{rt} + b_4 \text{Log } F_t + u_t$$

$$\text{A.4 } \text{Log } A_{rt} = a + b_1 \text{Log } R_{rt} + b_2 \text{ Log } I_t + u_t$$

Where $A_{rt}$, $P_{rt}$, $Y_{rt}$ and $G_{rt}$ are respectively areas, farm harvest price, yield rate and gross return of the rth crop in year t. $R_{rt}$ stands for the pre-sowing rainfall relevant to the rth crop, $F_t$ for the index of fertiliser price and $I_t$ for the total irrigated area during the rabi season. $u_t$ is the error term.

Equation A.1 and A.2 are used for estimating the acreage response of winter rice, autumn rice, maize, ragi and barley. The two additional specifications i.e. A.3 and A.4 are used for the wheat crop because irrigation and fertiliser price are two important considerations in the growing of wheat. Equation A.3 is estimated only for the HYV period because the chemical fertilisers have been used in growing wheat in the HYV period.

### 5.2.3 **Regression Results**

We have estimated the regression equations for each crop and for each district and for Chotanagpur region (pooled) for the two sub-periods and for the combined period.[2] Regression results are presented in Table 5.1 to 5.6. We have also computed the F-ratios associated with the test of overall homogeneity of regressions in the two sub-periods and the results are given in Table 5.7. We consider the regression results for kharif cereals first and then those for rabi cereals.

---

2. It should be noted that t and F tests are only asymptotically valid because of the presence of the lagged endogenous variable. For this reason the combined period and the region results are more reliable than the individual period and individual district results.

TABLE 5.1: **Showing acreage response equations estimated for individual crops in the two sub-periods and combined period - Winter rice**

| Constant, Independent | CHOTANAGPUR REGION AS A WHOLE (PANEL DATA) Pre-HYV period | | HYV period | | Combined period | |
|---|---|---|---|---|---|---|
| Variables, $R^2$ and F | Eqn. No. A.1 | Eqn. No. A.2 | Eqn. No. A.1 | Eqn. No. A.2 | Eqn.No. A.1 | Eqn. No. A.2 |
| 1 | 2 | 3 | 4 | 5 | 6 | 7 |
| 1. Constant | 0.110 | 0.044 | -0.986 | -1.164 | -0.450 | -0.760 |
| 2. Log $A_{rt-1}$ | 0.973*** (0.014) | 0.974*** (0.015) | 1.000*** (0.019) | 1.008*** (0.019) | 0.991*** (0.011) | 0.989*** (0.012) |
| 3. Log $P_{rt-1}$ | 0.005 (0.049) | - | 0.038 (0.046) | - | -0.007 (0.020) | - |
| 4. Log $Y_{rt-1}$ | -0.089* (0.034) | - | -0.082 (0.049) | - | -0.080** (0.028) | - |
| 5. Log $G_{rt-1}$ | - | -0.067* (0.035) | - | -0.019 (0.031) | - | -0.031* (0.017) |
| 6. Log $R_{rt}$ | 0.103 (0.062 | 0.084 (0.063) | 0.197** (0.057) | 0.185** (0.057) | 0.164*** (0.041) | 0.156*** (0.041) |
| 7. $R^2$ | 0.988 | 0.988 | 0.982 | 0.980 | 0.982 | 0.982 |
| 8. $F_{V1, V2)}$ | 1172.1*** (4,55) | 1470.7*** (3,56) | 734.4*** (4,55) | 948.6*** (3,56) | 1761.5*** (4,115) | 2268.1*** (3,116) |
| | SANTHAL PARGANA DISTRICT | | | | | |
| 1. | 9.200 | 8.899 | 11.206 | 11.979 | 10.270 | 10.120 |
| 2. | -0.307 (0.310) | -0.320 (0.299) | -0.676 (0.348) | -0.653 (0.312) | -0.337 (0.214) | -0.336 (0.201) |
| 3. | 0.037 (0.087) | - | 0.121 (0.131) | - | -0.033 (0.042) | - |
| 4. | -0.052 (0.140) | - | 0.201 (0.250) | - | -0.030 (0.126) | - |
| 5. | - (0.081) | 0.027 | - (0.110) | 0.137 | - (0.040) | -0.032 |
| 6. | 0.300* (0.146) | 0.305* (0.141) | 0.134 (0.110) | 0.141 (0.099) | 0.191** (0.081) | 0.192** (0.077) |

*Contd.* Table 5.1

| 1 | 2 | 3 | 4 | 5 | 6 | 7 |
|---|---|---|---|---|---|---|
| 7. | 0.540 | 0.484 | 0.561 | 0.554 | 0.356 | 0.356 |
| 8. | 1.469 | 1.883 | 1.601 | 2.486 | 2.080 | 2.960 |
| | (4,5) | (3,6) | (4,5) | (3,6) | (4,15) | (3,16) |
| | | | HAZARIBAGH DISTRICT | | | |
| 1. | 4.525 | 4.692 | 7.176 | 8.090 | 0.420 | 1.110 |
| 2. | 0.313 (0.392) | 0.279 (0.305) | -0.083 (0.458) | -0.206 (0.226) | 0.829*** (0.263) | 0.713*** (0.185) |
| 3. | -0.028 (0.085) | - | 0.058 (0.061) | - | -0.010 (0.056) | - |
| 4. | -0.042 (0.068) | - | -0.010 (0.167) | - | -0.074 (0.077) | - |
| 5. | - | -0.038 (0.054) | - | 0.041 (0.032) | - | -0.033 (0.042) |
| 6. | 0.195 (0.116) | 0.184* (0.085) | 0.150* (0.064) | 0.142** (0.054) | 0.218** (0.079) | 0.199** (0.072) |
| 7. | 0.559 | 0.558 | 0.686 | 0.79 | 0.658 | 0.650 |
| 8. | 1.589 (4,5) | 2.525 (3,6) | 2.733 (4,5) | 4.245* (3,6) | 7.230*** (4,15) | 9.890*** (3,16) |
| | | | PALAMAU DISTRICT | | | |
| 1. | 4.413 | 3.609 | 3.893 | 5.590 | 3.540 | 3.330 |
| 2. | 0.404*** (0.061) | 0.418*** (0.064) | 0.190 (0.413) | -0.007 (0.289) | 0.354*** (0.073) | 0.370*** (0.067) |
| 3. | -0.122 (0.074) | - | 0.053 (0.063) | - | -0.018 (0.022) | - |
| 4. | -0.049 (0.034) | - | -0.011 (0.040) | - | -0.001 (0.022) | - |
| 5. | - | -0.037 (0.034) | - | 0.011 (0.021) | - | -0.009 (0.016) |
| 6. | 0.064 (0.053) | 0.085 (0.053) | 0.217* (0.093) | 0.175** (0.068) | 0.137** (0.051) | 0.149*** (0.045) |

*Contd.* Table 5.1

| 1 | 2 | 3 | 4 | 5 | 6 | 7 |
|---|---|---|---|---|---|---|
| 7. | 0.925 | 0.901 | 0.608 | 0.570 | 0.699 | 0.691 |
| 8. | 15.43***<br>(4,5) | 18.05***<br>(3,6) | 1.940<br>(4,5) | 2.646<br>(3,6) | 8.71***<br>(4,15) | 11.97***<br>(3,16) |
| | | | RANCHI DISTRICT | | | |
| 1. | 15.319 | 16.346 | 11.176 | 11.530 | 12.010 | 12.220 |
| 2. | -1.009<br>(0.561) | -1.145**<br>(0.365) | -0.465<br>(0.239) | -0.473*<br>(0.211) | -0.544**<br>(0.202) | -0.558***<br>(0.172) |
| 3. | 0.032<br>(0.037) | - | 0.055*<br>(0.024) | - | 0.040***<br>(0.011) | - |
| 4. | 0.016<br>(0.042) | - | 0.062<br>(0.043) | - | 0.028<br>(0.023) | - |
| 5. | - | 0.059*<br>(0.028) | - | 0.057***<br>(0.014) | - | 0.043***<br>(0.008) |
| 6. | 0.123<br>(0.070) | 0.118*<br>(0.052) | 0.019<br>(0.033) | 0.019<br>(0.030) | 0.034<br>(0.028) | 0.033<br>(0.024) |
| 7. | 0.475 | 0.651 | 0.757 | 0.755 | 0.532 | 0.621 |
| 8. | 1.133<br>(4,5) | 3.724*<br>(3,6) | 3.896*<br>(4,5) | 6.212**<br>(3,6) | 4.270**<br>(4,15) | 8.730***<br>(3,16) |
| | | | SINGHBHUM DISTRICT | | | |
| 1. | 6.387 | 3.162 | 11.277 | 11.744 | 5.620 | 6.500 |
| 2. | 0.383<br>(0.498) | 0.804<br>(0.485) | -0.314<br>(0.288) | -0.349<br>(0.264) | 0.377*<br>(0.199) | 0.306<br>(0.200) |
| 3. | -0.072<br>(0.107) | - | -0.002<br>(0.033) | - | 0.027<br>(0.021) | - |
| 4. | 0.062<br>(0.067) | - | 0.060<br>(0.098) | - | 0.091*<br>(0.044) | - |
| 5. | - | 0.066<br>(0.076) | - | 0.007<br>(0.027) | - | 0.034<br>(0.021) |
| 6. | -0.233<br>(0.125) | -0.297*<br>(0.135) | -0.152*<br>(0.066) | -0.127*<br>(0.047) | -0.198***<br>(0.055) | -0.166***<br>(0.052) |

*Contd.* Table 5.1

| 1 | 2 | 3 | 4 | 5 | 6 | 7 |
|---|---|---|---|---|---|---|
| 7. | 0.645 | 0.452 | 0.638 | 0.616 | 0.503 | 0.433 |
| 8. | 2.272<br>(4,5) | 1.647<br>(3,6) | 2.207<br>(4,5) | 3.205<br>(3,6) | 3.800**<br>(4,15) | 4.080**<br>(3,16) |
| | | | DHANBAD DISTRICT | | | |
| 1. | 0.701 | 1.282 | 8.700 | 8.071 | 3.660 | 2.422 |
| 2. | 0.300<br>(0.383) | 0.228<br>(0.265) | -0.543<br>(0.603) | -0.458<br>(0.501) | 0.322<br>(0.366) | 0.467*<br>(0.265) |
| 3. | 0.046<br>(0.103) | - | -0.120<br>(0.121) | - | -0.170*<br>(0.085) | - |
| 4. | 0.007<br>(0.123) | - | -0.024<br>(0.239) | - | -0.065<br>(0.145) | |
| 5. | - | 0.031<br>(0.081) | - | -0.095<br>(0.086) | - | -0.140*<br>(0.066) |
| 6. | 0.575**<br>(0.187) | 0.563**<br>(0.168) | 0.312*<br>(0.132) | 0.316**<br>(0.121) | 0.293**<br>(0.112) | 0.299**<br>(0.109) |
| 7. | 0.669 | 0.664 | 0.584 | 0.575 | 0.560 | 0.551 |
| 8. | 2.526<br>(4,5) | 3.949*<br>(3,6) | 1.752<br>(4,5) | 2.706<br>(3,6) | 3.470**<br>(4,15) | 6.540***<br>(3,16) |

Regression results for kharif cereals viz. winter rice, autumn rice, maize and ragi are given in Table 5.1 to 5.4 respectively. $R^2$'s are very high in the regression for the Chotanagpur region as a whole estimated from the panel data. The coefficients of the lagged dependent variable in the equation for the Chotanagpur region are highly significant in all the cases and approximately equal to 1 indicating that almost the same amount of land under these crops has been maintained. We find insignificant price coefficients in all the cases, however, the absolute values of the price co-efficients have increased in the HYV period as compared to those in the pre-HYV period in the case of winter rice and ragi. Yield rate coefficients are negative in all the cases and significant at 5 per cent level of significance in the combined period regression in the case of winter rice and ragi. This might be so because of the subsistence characteristics of the crops as the farmers may have maintained the area under these crops despite a falling trend in the yield rate of these crops at the Chotanagpur level in the HYV period. The coefficients of pre-sowing rainfall variable are positive and in most cases significant for winter rice and autumn rice but negative and in most cases significant for maize and ragi. This may be so because maize and ragi require less water than other cereals like autumn rice grown in the kharif season and the heavy rainfall during the sowing season may have adversely affected the area under these crops favouring in their place the other cereals.

TABLE 5.2 : **Showing acreage response equations estimated for individual crops in the two sub-periods and combined period - Autumn rice**

| Constant, Independent Variables, $R^2$ and F | CHOTANAGPUR REGIONS AS A WHOLE (PANEL DATA) | | | | | |
|---|---|---|---|---|---|---|
| | Pre-HYV period | | HYV period | | Combined period | |
| | Eqn. No. A.1 | Eqn. No. A.2 | Eqn. No. A.1 | Eqn. No. A.2 | Eqn. No. A.1 | Eqn. No. A.2 |
| 1 | 2 | 3 | 4 | 5 | 6 | 7 |
| 1. Constant | 0.164 | -0.330 | 0.706 | -0.026 | 0.464 | -0.269 |
| 2. Log $A_{rt-1}$ | 0.973***<br>(0.026) | 0.976***<br>(0.026) | 0.972***<br>(0.030) | 0.979***<br>(0.029) | 0.971***<br>(0.019) | 0.979***<br>(0.019) |
| 3. Log $P_{rt-1}$ | 0.040<br>(0.083) | - | -0.015<br>(0.089) | - | 0.030<br>(0.037) | - |
| 4. Log $Y_{rt-1}$ | -0.116<br>(0.118) | - | -0.154<br>(0.105) | - | -0.138*<br>(0.074) | - |
| 5. Log $G_{rt-1}$ | - | -0.006<br>(0.069) | - | -0.072<br>(0.066) | - | -0.007<br>(0.032) |
| 6. Log $R_{rt}$ | 0.103<br>(0.116) | 0.095<br>(0.118) | 0.094<br>(0.096) | 0.107<br>(0.095) | 0.084<br>(0.068) | 0.082<br>(0.069) |
| 7. $R^2$ | 0.972 | 0.970 | 0.964 | 0.962 | 0.987 | 0.966 |
| 8. F<br>$(v_1,v_2)$ | 386.6***<br>(4,45) | 512.5***<br>(3,46) | 298.3***<br>(4,45) | 397.4***<br>(3,46) | 700.9***<br>(4,95) | 907.6***<br>(3,96) |
| | SANTHAL PARGANA DISTRICT | | | | | |
| 1. | 6.703 | 7.924 | 7.942 | 5.787 | 5.573 | 3.649 |
| 2. | 0.214<br>(0.447) | 0.148<br>(0.363) | 0.036<br>(0.303) | 0.033<br>(0.296) | 0.555**<br>(0.226) | 0.492**<br>(0.221) |
| 3. | 0.274 | - | 0.352 | - | 0.060 | - |

| | | | | | | |
|---|---|---|---|---|---|---|
| | (0.321) | | (0.191) | | (0.146) | |
| 4. | 0.072 | - | -0.221 | - | -0.452 | - |
| | (0.406) | | (0.541) | | (0.335) | |
| 5. | - | 0.187 | - | 0.272 | - | -0.074 |
| | | (0.146) | | (0.162) | | (0.089) |
| 6. | -0.105 | -0.125 | 0.114 | 0.214 | 0.182 | 0.234 |
| | (0.230) | (0.204) | (0.183) | (0.145) | (0.144) | (0.139) |
| 7. | 0.308 | 0.293 | 0.596 | 0.539 | 0.526 | 0.483 |
| 1 | 2 | 3 | 4 | 5 | 6 | 7 |
| 8. | 0.557 | 0.828 | 1.840 | 2.333 | 4.170** | 4.983** |
| | (4,5) | (3,6) | (4,5) | (3,6) | (4,15) | (3,16) |
| **HAZARIBAGH DISTRICT** | | | | | | |
| 1. | 9.196 | 8.385 | 14.546 | 15.101 | 10.836 | 10.867 |
| 2. | -0.097 | -0.011 | -0.719 | -0.755 | -0.160 | -0.222 |
| | (0.545) | (0.386) | (0.880) | (0.794) | (0.350) | (0.325) |
| 3. | -0.131 | - | 0.104 | - | 0.082 | - |
| | (0.161) | | (0.371) | | (0.116) | |
| 4. | -0.002 | - | 0.078 | - | -0.053 | - |
| | (0.480) | | (0.403) | | (0.260) | |
| 5. | - | -0.119 | - | 0.106 | - | 0.072 |
| | | (0.137) | | (0.303) | | (0.111) |
| 6. | 0.189 | 0.213 | 0.044 | 0.058 | -0.051 | -0.025 |
| | (0.298) | (0.261) | (0.280) | (0.237) | (0.161) | (0.150) |
| 7. | 0.234 | 0.229 | 0.309 | 0.311 | 0.058 | 0.043 |
| 8. | 0.382 | 0.594 | 0.559 | 0.906 | 0.235 | 0.237 |
| | (4,5) | (3,6) | (4,5) | (3,6) | (4,15) | (3,16) |
| **PALAMAU DISTRICT** | | | | | | |
| 1. | -2.471 | 1.523 | 4.700 | 4.807 | 1.637 | 1.807 |
| 2. | 0.874** | 0.682** | 0.248 | 0.265 | 0.636*** | 0.645*** |
| | (0.241) | (0.173) | (0.298) | (0.275) | (0.151) | (0.142) |
| 3. | 0.400 | - | 0.103 | - | 0.028 | - |
| | (0.240) | | (0.084) | | (0.054) | |
| 4. | 0.054 | - | 0.055 | - | 0.056 | - |
| | (0.195 | | (0.092) | | (0.090) | |

*Contd.* Table 5.2

| 1 | 2 | 3 | 4 | 5 | 6 | 7 |
|---|---|---|---|---|---|---|
| 5. | - | 0.138<br>(0.126) | - | 0.081<br>(0.061) | - | 0.034<br>(0.045) |
| 6. | 0.310<br>(0.232) | 0.140<br>(0.203) | 0.278*<br>(0.129) | 0.285*<br>(0.119) | 0.228*<br>(0.111) | 0.229*<br>(0.108) |
| 7. | 0.812 | 0.741 | 0.594 | 0.581 | 0.585 | 0.582 |
| 8. | 0.400**<br>(4,5) | 5.740**<br>(3,6) | 1.838<br>(4,5) | 2.769<br>(3,6) | 5.286***<br>(4,15) | 7.422***<br>(3,16) |
| | | | **RANCHI DISTRICT** | | | |
| 1. | 12.877 | 13.669 | 8.637 | 9.012 | 6.134 | 5.098 |
| 2. | -0.167<br>(0.400) | -0.166<br>(0.386) | 0.228<br>(0.347) | 0.180<br>(0.324) | 0.406*<br>(0.219) | 0.509**<br>(0.211) |
| 3. | 0.037<br>(0.058) | - | 0.045<br>(0.059) | - | 0.093**<br>(0.036) | - |
| 4. | 0.125<br>(0.188) | - | -0.043<br>(0.118) | - | 0.033<br>(0.106) | - |
| 5. | - | 0.006<br>(0.055) | - | 0.027<br>(0.050) | - | 0.075**<br>(0.035) |
| 6. | -0.037<br>(0.129) | -0.030<br>(0.131) | 0.079<br>(0.085) | 0.075<br>(0.080) | 0.034<br>(0.073) | 0.035<br>(0.073) |
| 7. | 0.153 | 0.039 | 0.365 | 0.308 | 0.694 | 0.655 |
| 8. | 0.226<br>(4,5) | 0.081<br>(3,6) | 0.719<br>(4,5) | 0.891<br>(3,6) | 8.538***<br>(4,15) | 10.13***<br>(3,16) |
| | | | **SINGHBHUM DISTRICT** | | | |
| 1. | 0.458 | -0.181 | 4.034 | 5.559 | 1.503 | 1.304 |
| 2. | 0.791*<br>(0.334) | 0.806*<br>(0.299) | 0.452*<br>(0.168) | 0.395*<br>(0.166) | 0.763***<br>(0.159) | 0.742***<br>(0.151) |
| 3. | 0.093<br>(0.421) | - | 0.057<br>(0.051) | - | 0.069<br>(0.084) | - |
| 4. | -0.077<br>(0.445) | - | 0.212**<br>(0.097) | - | -0.066<br>(0.182) | - |

*Contd.* Table 5.2

| 1 | 2 | 3 | 4 | 5 | 6 | 7 |
|---|---|---|---|---|---|---|
| 5. | - | 0.011<br>(0.177) | - | 0.104**<br>(0.032) | - | 0.034<br>(0.056) |
| 6. | 0.272<br>(0.479) | 0.323<br>(0.385) | -0.017<br>(0.064) | -0.025<br>(0.066) | 0.154<br>(0.168) | 0.172<br>(0.161) |
| 7. | 0.584 | 0.579 | 0.902 | 0.876 | 0.674 | 0.667 |
| 8. | 1.748<br>(4,5) | 2.751<br>(3,6) | 11.610***<br>(4,5) | 14.090***<br>(3,6) | 7.767***<br>(4,15) | 10.691***<br>(3,16) |

Regression equations estimated for individual districts, however, indicate significantly positive coefficients of price and gross return variables in the case of Ranchi district for all the four crops in the combined period regression, in the Palamau for maize in the combined period regression, in Dhanbad for maize in the pre-HYV period and in Singhbhum for autumn rice in the HYV period. The lagged dependent variable in the regression of some districts (specially in the regression equation for winter rice in Ranchi we have already noted that the tribal concentration is highest in the Ranchi district) are found to be negative and significant. This may be so because of the "traditional" propensity of the tribal farmers to maintain a uniform amount of land under cultivation over the years so that if for any reason less land is brought under cultivation in one year, this is made up by bringing a large amount of land in the next year and vice versa. This behaviour may also be treated as a facet of traditional behaviour. In fact the total area under cultivation is limited but there is flexibility because of the presence of culturable waste land including fallow lands and the forest land which because of gradual deforestation is becoming available for cultivation. Unexpectedly the coefficient of the pre-sowing rainfall variable is significantly negative in the case of winter rice for Singhbhum district. This may be due to the water logging difficulties which restricts the area available for cultivation during the heavy rainfall years.

The overall homogeneity of regressions in the two sub-periods is accepted in all these cases except in the case of winter rice for Hazaribagh, autumn rice for Santhal Pargana, and maize for Ranchi (details are given in Table 5.7). It should be noted that the $R^2$ 's of the regression equations estimated for the individual districts in the case of winter rice which is the major cereal and which has practically no substitute crop in the season and on the land in which it is grown, is not very high. In the case of the minor cereals, however, the $R^2$'s are very small, which may be due to their possible substitution by other crops and substitution among themselves. The crop-substitution issue is examined in the next chapter.

TABLE 5.3 : **Showing acreage response equations estimated for individual crops in the two sub-periods and combined period - Maize**

| Constant, Independent Variables, $R^2$ and F | CHOTANAGPUR REGION AS A WHOLE (PANEL DATA) | | | | | |
|---|---|---|---|---|---|---|
| | Pre-HYV period | | HYV period | | Combined period | |
| | Eqn. No. A.1 | Eqn. No. A.2 | Eqn. No. A.1 | Eqn. No. A.2 | Eqn. No. A.1 | Eqn. No. A.2 |
| 1 | 2 | 3 | 4 | 5 | 6 | 7 |
| 1. Constant | 1.382 | 1.225 | 2.314 | 1.950 | 1.789 | 1.651 |
| 2. Log $A_{rt-1}$ | 0.993*** (0.014) | 0.990*** (0.013) | 0.969*** (0.028) | 0.973*** (0.027) | 0.984*** (0.014) | 0.984*** (0.014) |
| 3. Log $P_{rt-1}$ | -0.030 (0.044) | - | -0.095 (0.066) | - | -0.022 (0.022) | - |
| 4. Log $Y_{rt-1}$ | -0.063* (0.030) | - | -0.019 (0.056) | - | -0.040 (0.032) | - |
| 5. Log $G_{rt-1}$ | - | -0.052* (0.025) | - | -0.043 (0.043) | - | -0.026 (0.017) |
| 6. Log $R_{rt}$ | -0.125** (0.058) | -0.132* (0.057) | -0.231*** (0.082) | -0.222** (0.079) | -0.203*** (0.050) | -0.211*** (0.047) |
| 7. $R^2$ | 0.990 | 0.990 | 0.958 | 0.958 | 0.974 | 0.974 |
| 8. F $(v_1,v_2)$ | 1463.9*** (4,55) | 1973.0*** (3,56) | 323.19*** (4,55) | 429.18*** (3,56) | 1120.3*** (4,115) | 1499.0*** (3,116) |
| | SANTHAL PARGANA DISTRICT | | | | | |
| 1. | 2.825 | 2.697 | 2.555 | 2.887 | 11.049 | 9.132 |
| 2. | 0.830 (0.491) | 0.825 (0.458) | -1.900 (1.053) | -1.914* (0.952) | -0.007 (0.363) | 0.210 (0.302) |
| 3. | -0.039 (0.126) | - | 0.689 (0.404) | - | -0.013 (0.063) | - |

*Contd. Table 5.3*

| 1 | 2 | 3 | 4 | 5 | 6 | 7 |
|---|---|---|---|---|---|---|
| 4. | -0.022<br>(0.081) | - | 0.700*<br>(0.340) | - | 0.109<br>(0.108) | - |
| 5. | - | -0.024<br>0.072) | - | 0.703*<br>(0.310) | - | 0.010<br>(0.058) |
| 6. | -0.107<br>(0.133) | -0.102<br>(0.116) | -0.253<br>(0.198) | -0.254<br>(0.181) | -0.111<br>(0.126) | -0.087<br>(0.125) |
| 7. | 0.524 | 0.521 | 0.471 | 0.472 | 0.128 | 0.062 |
| 8. | 1.377<br>(4,5) | 2.181<br>(3,6) | 1.113<br>(4,5) | 1.789<br>(3,6) | 0.553<br>(4,15) | 0.349<br>(3,16) |
| | | | HAZARIBAGH DISTRICT | | | |
| 1. | 8.834 | 7.684 | 13.712 | 13.971 | 7.971 | 8.284 |
| 2. | 0.271<br>(0.427) | 0.369<br>(0.372) | -0.217<br>(0.518) | -0.284<br>(0.492) | 0.331<br>(0.309) | 0.309<br>(0.274) |
| 3. | -0.083<br>(0.102) | - | -0.086<br>(0.133) | - | 0.062<br>(0.065) | - |
| 4. | -0.001<br>(0.069) | - | 0.056<br>(0.119) | - | 0.014<br>(0.069) | - |
| 5. | - | -0.030<br>(0.047) | - | 0.036<br>(0.089) | - | 0.061<br>(0.046) |
| 6. | -0.132<br>(0.118) | -0.136<br>(0.111) | -0.097<br>(0.190) | -0.061<br>(0.171) | -0.189<br>(0.112) | -0.206*<br>(0.095) |
| 7. | 0.308 | 0.260 | 0.271 | 0.193 | 0.358 | 0.384 |
| 8. | 0.557<br>(4,5) | 0.701<br>(3,6) | 0.465<br>(4,5) | 0.477<br>(3,6) | 2.089<br>(4,15) | 3.332**<br>(3,16) |
| | | | PALAMAU DISTRICT | | | |
| 1. | 9.982 | 10.246 | 7.956 | 8.278 | 6.675 | 6.318 |
| 2. | 0.154<br>(0.233) | 0.145<br>(0.219) | 0.290<br>(0.171) | 0.287<br>(0.151) | 0.430***<br>(0.127) | 0.514***<br>(0.123) |
| 3. | 0.036<br>(0.085) | - | 0.061<br>(0.036) | - | 0.121***<br>(0.039) | - |

*Contd. Table 5.3*

| 1 | 2 | 3 | 4 | 5 | 6 | 7 |
|---|---|---|---|---|---|---|
| 4. | -0.016<br>(0.064) | - | 0.065<br>(0.035) | - | 0.034<br>(0.039) | - |
| 5. | - | 0.004<br>(0.055) | - | 0.063**<br>(0.030) | - | 0.078**<br>(0.030) |
| 6. | -0.181<br>(0.108) | -0.206*<br>(0.094) | -0.163<br>(0.070) | -0.161**<br>(0.062) | -0.211***<br>(0.068) | -0.259***<br>(0.065) |
| 7. | 0.501 | 0.468 | 0.671 | 0.671 | 0.880 | 0.859 |
| 8. | 1.254<br>(4,5) | 1.755<br>(3,6) | 2.551<br>(4,5) | 4.075*<br>(3,6) | 27.67***<br>(4,15) | 32.47***<br>(3,16) |
| | | | RANCHI DISTRICT | | | |
| 1. | 8.609 | 7.045 | 12.079 | 12.522 | 7.227 | 6.213 |
| 2. | 0.193<br>(0.662) | 0.307<br>(0.598) | -0.159<br>(0.225) | -0.158<br>(0.187) | 0.250<br>(0.243) | 0.530**<br>(0.185) |
| 3. | -0.142<br>(0.201) | - | 0.098<br>(0.118) | - | 0.320**<br>(0.117) | - |
| 4. | -0.031<br>(0.110) | - | 0.097<br>(0.080) | - | 0.108<br>(0.084) | |
| 5. | - | -0.042<br>(0.102) | - | 0.097<br>(0.065) | - | 0.169**<br>(0.080) |
| 6. | -0.086<br>(0.174) | -0.079<br>(0.164) | -0.321<br>(0.173) | -0.321*<br>(0.154) | -0.348*<br>(0.180) | -0.446**<br>(0.180) |
| 7. | 0.121 | 0.060 | 0.551 | 0.552 | 0.790 | 0.752 |
| 8. | 0.172<br>(4,5) | 0.128<br>(3,6) | 1.535<br>(4,5) | 2.457<br>(3,6) | 14.236***<br>(4,15) | 16.228***<br>(3,16) |
| | | | SINGHBHUM DISTRICT | | | |
| 1. | 7.201 | 7.452 | -0.943 | 3.025 | 5.168 | 4.384 |
| 2. | 0.346<br>(0.436) | 0.348<br>(0.400) | 1.126*<br>(0.513) | 0.720*<br>(0.301) | 0.507**<br>(0.198) | 0.592***<br>(0.178) |
| 3. | 0.009<br>(0.158) | - | 0.052<br>(0.043) | - | -0.049<br>(0.036) | - |

*Contd. Table 5.3*

| 1 | 2 | 3 | 4 | 5 | 6 | 7 |
|---|---|---|---|---|---|---|
| 4. | 0.047<br>(0.153) | - | -0.025<br>(0.057) | - | 0.020<br>(0.051) | - |
| 5. | - | 0.028<br>(0.104) | - | 0.022<br>(0.030) | - | -0.021<br>(0.020) |
| 6. | -0.214<br>(0.209) | -0.228<br>(0.178) | -0.050<br>(0.108) | -0.089<br>(0.100) | -0.088<br>(0.086) | -0.065<br>(0.081) |
| 7. | 0.475 | 0.472 | 0.584 | 0.503 | 0.450 | 0.421 |
| 8. | 1.131<br>(4,5) | 1.785<br>(3,6) | 1.747<br>(4,5) | 2.024<br>(3,6) | 3.079**<br>(4,15) | 3.886**<br>(3,16) |
| | | | DHANBAD DISTRICT | | | |
| 1. | 8.863 | 10.354 | 9.456 | 9.692 | 6.657 | 2.156 |
| 2. | -0.448<br>(0.308) | -0.490<br>(0.274) | -0.056<br>(0.639) | -0.120<br>(0.537) | 0.180<br>(0.346) | 0.222<br>(0.353) |
| 3. | 0.266**<br>(0.079) | - | -0.028<br>(0.176) | - | 0.133<br>(0.079) | - |
| 4. | 0.211*<br>(0.084) | - | -0.117<br>(0.254) | - | -0.044<br>(0.115) | - |
| 5. | - | 0.239***<br>(0.051) | - | -0.061<br>(0.103) | - | 0.082<br>(0.071) |
| 6. | 0.196<br>(0.144) | 0.179<br>(0.123) | 0.116<br>(0.229) | 0.087<br>(0.180) | 0.034<br>(0.144) | -0.070<br>(0.126) |
| 7. | 0.835 | 0.832 | 0.099 | 0.089 | 0.371 | 0.292 |
| 8. | 6.320**<br>(4,5) | 9.943***<br>(3,6) | 0.138<br>(4,5) | 0.176<br>(3,6) | 2.214<br>(4,15) | 2.196<br>(3,16) |

Regression results for rabi cereals viz. wheat and barley are presented in Table 5.5 and 5.6 respectively. $R^2$'s of the regressions estimated for the Chotanagpur region as a whole on the basis of panel data are very high. $R^2$'s in the regressions for the individual districts are also quite high for the wheat crop but not so for the barley crop. This may be because of the farmers' preference for wheat as compared to barley.

**TABLE 5.4 : Showing acreage response equations estimated for individual crops in the two sub-periods and combined period - Ragi**

| Constant, Independent Variables, $R^2$ and F | Pre-HYV period Eqn. No. A.1 | Pre-HYV period Eqn. No. A.2 | HYV period Eqn. No. A.1 | HYV period Eqn. No. A.2 | Combined period Eqn. No. A.1 | Combined period Eqn. No. A.2 |
|---|---|---|---|---|---|---|
| 1 | 2 | 3 | 4 | 5 | 6 | 7 |
| | **CHOTANAGPUR REGION AS A WHOLE (PANEL DATA)** | | | | | |
| 1. Constant | 1.181 | 1.987 | 1.930 | 1.496 | 2.093 | 1.593 |
| 2. Log $A_{rt-1}$ | 0.999*** (0.035) | 1.000*** (0.034) | 0.995*** (0.022) | 1.001*** (0.022) | 1.000*** (0.019) | 1.000*** (0.019) |
| 3. Log $P_{rt-1}$ | 0.028 (0.157) | - | 0.039 (0.066) | - | -0.020 (0.038) | - |
| 4. Log $Y_{rt-1}$ | -0.092 (0.107) | - | -0.151* (0.065) | - | -0.118** (0.059) | |
| 5. Log $G_{rt-1}$ | - | -0.054 (0.087) | - | -0.058 (0.048) | - | -0.048 (0.033) |
| 6. Log $R_{rt}$ | -0.270 (0.194) | -0.272 (0.195) | -0.183* (0.078) | -0.189* (0.081) | -0.206** (0.082) | -0.212** (0.082) |
| 7. $R^2$ | 0.953 | 0.951 | 0.979 | 0.978 | 0.966 | 0.965 |
| 8. F $(v_1,v_2)$ | 221.8*** (4,45) | 299.6*** (3,46) | 546.44*** (4,45) | 678.2*** (3,46) | 672.8*** (4,95) | 886.6*** (3,96) |
| | **SANTHAL PARGANA DISTRICT** | | | | | |
| 1. | 8.803 | 8.474 | 3.790 | 2.937 | 3.933 | 3.786 |
| 2. | 0.132 (0.408) | 0.150 (0.377) | 0.992 (0.655) | 0.791 (0.633) | 0.697*** (0.215) | 0.648*** (0.188) |
| 3. | -0.022 (0.209) | - | -0.089 (0.369) | - | -0.061 (0.072) | - |
| 4. | 0.079 (0.121) | - | -0.541 (0.488) | - | -0.148 (0.128) | - |
| 5. | - | 0.059 (0.103) | - | -0.189 (0.361) | - | -0.093 (0.065) |
| 6. | -0.316 (0.232) | -0.268 (0.169) | -0.036 (0.231) | -0.042 (0.234) | -0.061 (0.148) | -0.075 (0.143) |
| 7. | 0.310 | 0.285 | 0.412 | 0.279 | 0.494 | 0.486 |
| 8. | 0.563 (4,5) | 0.799 (3,6) | 0.878 (4,5) | 0.775 (3,6) | 3.665** (4,15) | 5.040*** (3,16) |

*Contd. Table 5.4*

| 1 | 2 | 3 | 4 | 5 | 6 | 7 |
|---|---|---|---|---|---|---|
| | | | HAZARIBAGH DISTRICT | | | |
| 1. | 7.120 | 6.670 | 10.904 | 10.344 | 9.621 | 8.703 |
| 2. | 0.372<br>(0.305) | 0.449<br>(0.285) | -0.023<br>(0.534) | 0.083<br>(0.464) | 0.125<br>(0.290) | 0.272<br>(0.260) |
| 3. | -0.015<br>(0.058) | - | 0.021<br>(0.132) | - | 0.005<br>(0.039) | - |
| 4. | 0.047<br>(0.055) | - | 0.150<br>(0.168) | - | 0.091<br>(0.070) | - |
| 5. | - | 0.018<br>(0.043) | - | 0.074<br>(0.079) | - | 0.023<br>(0.036) |
| 6. | -0.168<br>(0.096) | -0.195*<br>(0.088) | -0.262<br>(0.187) | -0.251<br>(0.174) | -0.225**<br>(0.102) | -0.243**<br>(0.101) |
| 7. | 0.635 | 0.581 | 0.394 | 0.361 | 0.357 | 0.305 |
| 8. | 2.179<br>(4,5) | 2.776<br>(3,6) | 0.824<br>(4,5) | 1.129<br>(3,6) | 2.001<br>(4,15) | 2.339<br>(3,16) |
| | | | PALAMAU DISTRICT | | | |
| 1. | 9.479 | 8.620 | 5.137 | 3.840 | 6.561 | 5.224 |
| 2. | 0.244<br>(0.346) | 0.170<br>(0.362) | 0.329<br>(0.271) | 0.378<br>(0.310) | 0.346<br>(0.217) | 0.357<br>(0.230) |
| 3. | -0.060<br>(0.238) | - | 0.248*<br>(0.099) | - | 0.058<br>(0.046) | - |
| 4. | -0.326<br>(0.249) | - | -0.053<br>(0.166) | - | -0.165<br>(0.124) | - |
| 5. | - | -0.174<br>(0.233) | - | 0.184<br>(0.106) | - | 0.044<br>(0.048) |
| 6. | -0.170<br>(0.149) | -0.149<br>(0.157) | -0.048<br>(0.181) | -0.048<br>(0.198) | -0.058<br>(0.123) | -0.018<br>(0.128) |
| 7. | 0.426 | 0.228 | 0.658 | 0.454 | 0.320 | 0.182 |
| 8. | 0.931<br>(4,5) | 0.589<br>(3,6) | 2.410<br>(4,5) | 1.667<br>(3,6) | 1.769<br>(4,15) | 1.184<br>(3,16) |
| | | | RANCHI DISTRICT | | | |
| 1. | 9.143 | 10.071 | 4.096 | 4.405 | 10.100 | 10.480 |
| 2. | -0.040<br>(0.135) | -0.034<br>(0.122) | 0.596<br>(0.379) | 0.617<br>(0.437) | -0.021<br>(0.080) | -0.020<br>(0.078) |

*Contd. Table 5.4*

| 1 | 2 | 3 | 4 | 5 | 6 | 7 |
|---|---|---|---|---|---|---|
| 3. | 0.090 | - | 0.140* | - | 0.109** | - |
| | (0.265) | | (0.071) | | (0.047) | |
| 4. | 0.182 | - | 0.035 | - | 0.076 | - |
| | (0.351) | | (0.059) | | (0.091) | |
| 5. | - | 0.119 | - | 0.072 | - | 0.106** |
| | | (0.221) | | (0.063) | | (0.045) |
| 6. | 0.101 | 0.077 | -0.082 | -0.104 | 0.010 | -0.002 |
| | (0.450) | (0.406) | (0.093) | (0.106) | (0.145) | (0.137) |
| 7. | 0.082 | 0.072 | 0.714 | 0.545 | 0.273 | 0.266 |
| 8. | 0.113 | 0.155 | 3.129 | 2.394 | 1.408 | 1.937 |
| | (4,5) | (3,6) | (4,5) | (3,6) | (4,15) | (3,16) |
| | | | DHANBAD DISTRICT | | | |
| 1. | 9.053 | 8.450 | 5.385 | 4.580 | 6.419 | 6.383 |
| 2. | -0.107 | 0.066 | 0.786 | 0.700 | 0.309 | 0.239 |
| | (0.466) | (0.476) | (0.771) | (0.541) | (0.335) | (0.312) |
| 3. | 0.481 | - | -0.243 | - | 0.029 | - |
| | (0.341) | | (0.189) | | (0.091) | |
| 4. | -0.211 | - | -0.295 | - | -0.076 | - |
| | (0.304) | | (0.269) | | (0.164) | |
| 5. | - | 0.104 | - | -0.251 | - | 0.016 |
| | | (0.202) | | 90.168) | | (0.087) |
| 6. | -0.109 | -0.270 | -0.144 | -0.119 | -0.102 | -0.081 |
| | (0.560) | (0.681) | (0.260) | (0.199) | (0.185) | (0.179) |
| 7. | 0.362 | 0.135 | 0.299 | 0.294 | 0.096 | 0.070 |
| 8. | 0.709 | 0.313 | 0.533 | 0.833 | 0.403 | 0.403 |
| | (4,5) | (3,6) | (4,5) | (3,6) | (4,15) | (3,16) |

Table 5.5 showing the regression results for wheat indicates that the farmers are responsive to yield rate and gross return rather than to prices, as indicated by the positive and significant coefficients of the yield rate and the gross return variables. The absolute values of the coefficients of the gross return variable increased in the HYV period as compared to those in the pre-HYV period in all the districts except in Palamau and Singhbhum where the use of HYV wheat is poor as compared to the other districts of Chotanagpur region. Irrigation facilities given to the farmers during the rabi season appear to constitute the most important factor influencing farmers' decision for planting wheat. In most cases the acreage elasticity with respect to irrigation exceeds the acreage elasticity with respect to gross return in the HYV period. It should be noted,

however, that the two elasticities are not directly comparable because the two variables do not appear in the same regression.[3]

---

3. Inclusion of irrigation variable with lagged dependent variable and gross return variable is found to lead to a change in the sign of the coefficients and also increases the values of the standard errors relative to the magnitudes of the estimated coefficients due to very high correlations between the irrigation variable ($I_t$)and the lagged dependent variable ($A_{t-1}$) and gross return variable ($G_{t-1}$) in the HYVperiod. The regression equations estimated for Ranchi and Singhbhum districts (for wheat crop)in the HYV period are given below:

Ranchi districts :

$$\text{(i)} \operatorname{Log} A_{rt} = -2.105 + \underset{(0.207)}{0.158} \operatorname{Log} A_{rt-1} + \underset{(0.254)}{1.005^{***}} \operatorname{Log} G_{rt-1} + \underset{(0.194)}{0.444} \operatorname{Log} R_{rt};$$

$$R^2 = 0.955,\ F_{(3,6)} = 43.73^{***}$$

$$\text{(ii)} \operatorname{Log} A_{rt} = -0.749 + \underset{(0.237)}{0.360} \operatorname{Log} A_{rt-1} + \underset{(0.437)}{1.007^{*}} \operatorname{Log} G_{rt-1} - \underset{(0.085)}{0.067} \operatorname{Log} I_t;$$

$$R^2 = 0.925,\ F_{(3,6)} = 25.00^{***}$$

$$\text{(iii)} \operatorname{Log} A_{rt} = 2.010 + \underset{(0.394)}{0.791} \operatorname{Log} R_{rt} + \underset{(0.078)}{0.286^{***}} \operatorname{Log} I_t;$$

$$R^2 = 0.701,\ F_{(2,7)} = 8.23^{**}$$

Singhbhum district :

$$\text{(i)} \operatorname{Log} A_{rt} = -4.867 + \underset{(0.136)}{0.543^{***}} \operatorname{Log} A_{rt-1} + \underset{(0.165)}{0.459^{**}} \operatorname{Log} G_{rt-1} + \underset{(0.225)}{0.945^{***}} \operatorname{Log} R_{rt};$$

$$R^2 = 0.935,\ F_{(3,6)} = 29.35^{***}$$

$$\text{(ii)} \operatorname{Log} A_{rt} = 2.405 - \underset{(0.123)}{0.156} \operatorname{Log} A_{rt-1} + \underset{(0.082)}{0.028} \operatorname{Log} G_{rt-1} + \underset{(0.107)}{0.923^{***}} \operatorname{Log} I_t;$$

$$R^2 = 0.980,\ F_{(3,6)} = 103.38^{***}$$

$$\text{(iii)} \operatorname{Log} A_{rt} = 1.186 + \underset{(0.111)}{0.166} \operatorname{Log} R_{rt} + \underset{(0.041)}{0.772^{***}} \operatorname{Log} I_t;$$

$$R^2 = 0.980,\ F_{(2,7)} = 189.80^{***}$$

TABLE 5.5: **Showing acreage response equations estimated for individual crops in the two sub-periods and combined period - Wheat**

| District/ region | Period | Eqn. No. | Constant | Coefficients of independent variables | | | | | | | $R^2$ | F $(v_1,v_2)$ |
|---|---|---|---|---|---|---|---|---|---|---|---|---|
| | | | | Log $A_{rt-1}$ | Log $P_{rt-1}$ | Log $Y_{rt-1}$ | Log $G_{rt-1}$ | Log $R_{rt}$ | Log $F_t$ | Log $I_t$ | | |
| 1 | 2 | 3 | 4 | 5 | 6 | 7 | 8 | 9 | 10 | 11 | 12 | 13 |
| Chotanagpur region as a whole (Panel Data) | Pre-HYV | A.1 | 0.017 | 0.929*** (0.041) | -0.630 (0.362) | 0.489** (0.172) | - | 0.003 (0.214) | - | - | 0.902 | 129.5*** (4,55) |
| | | A.2 | -2.005 | 0.923*** (0.043) | - | - | 0.233 (0.142) | 0.225 (0.204) | - | - | 0.893 | 156.2*** (3,56) |
| | | A.4 | -1.309 | - | - | - | - | 0.642* (0.344) | - | 0.851*** (0.078) | 0.672 | 58.68*** (2,57) |
| | HYV | A.1 | -2.333 | 0.894*** (0.038) | -0.059 (0.187) | 0.225* (0.122) | - | 0.372*** (0.127) | - | - | 0.916 | 154.1*** (4,55) |
| | | A.2 | -2.983 | 0.900*** (0.038) | - | - | 0.129 (0.086) | 0.361*** (0.127) | - | - | 0.916 | 204.5*** (3,56) |
| | | A.4 | 1.791 | - | - | - | - | 0.091 (0.229) | - | 0.780*** (0.064) | 0.719 | 73.16*** (2,57) |
| | Combined | A.1 | -2.791 | 0.914*** (0.028) | -0.005 (0.107) | 0.324*** (0.100) | - | 0.256** (0.114) | - | - | 0.916 | 316.3*** (4,115) |
| | | A.2 | -2.050 | 0.912*** (0.028) | - | - | 0.167*** (0.049) | 0.308*** (0.111) | - | - | 0.914 | 414.0*** (3,116) |

| 1 | 2 | 3 | 4 | 5 | 6 | 7 | 8 | 9 | 10 | 11 | 12 | 13 |
|---|---|---|---|---|---|---|---|---|---|---|---|---|
| | | A.4 | 0.639 | - | - | - | - | 0.421***<br>(0.209) | - | 0.710***<br>(0.043) | 0.699 | 136.4***<br>(2,117) |
| Santhal-Pargana | Pre-HYV | A.1 | 8.789 | -0.214<br>(0.250) | -0.133<br>(0.387) | 0.173<br>(0.111) | - | 0.149<br>(0.162) | - | - | 0.607 | 1.935<br>(4,5) |
| | | A.2 | 8.733 | -0.297<br>(0.187) | - | - | 0.162<br>(0.092) | 0.215<br>(0.125) | - | - | 0.615 | 3.193<br>(3,6) |
| | | A.4 | 6.508 | - | - | - | - | 0.252<br>(0.205) | - | 0.062<br>(0.287) | 0.213 | 0.950<br>(2,7) |
| | HYV | A.1 | -7.324 | 0.515*<br>(0.253) | 0.685<br>(0.488) | 0.722<br>(0.591) | - | 0.651<br>(0.428) | - | - | 0.922 | 14.72***<br>(4,5) |
| | | A.2 | -3.945 | 0.515*<br>(0.227) | - | - | 0.702<br>(0.427) | 0.636<br>(0.347) | - | - | 0.922 | 23.68***<br>(3,6) |
| | | A.3 | -3.731 | 0.417<br>(0.257) | - | - | 0.531<br>(0.477) | 0.617<br>(0.354) | 0.413<br>(0.468) | - | 0.933 | 17.29***<br>(4,5) |
| | | A.4 | 2.633 | - | - | - | - | 0.159<br>(0.258) | - | 0.660***<br>(0.059) | 0.947 | 63.16***<br>(3,6) |
| | Com-bined | A.1 | -2.162 | 0.568***<br>(0.157) | 0.438*<br>(0.242) | 0.349**<br>(0.153) | - | 0.320<br>(0.210) | - | - | 0.882 | 28.23***<br>(4,15) |
| | | A.2 | -0.548 | 0.576***<br>(0.128) | - | - | 0.400***<br>(0.107) | 0.318<br>(0.191) | - | - | 0.891 | 43.98***<br>(3,16) |
| | | A.4 | 2.349 | - | - | - | - | 0.421**<br>(0.155) | - | 0.525***<br>(0.061) | 0.925 | 106.4***<br>(2,17) |

| 1 | 2 | 3 | 4 | 5 | 6 | 7 | 8 | 9 | 10 | 11 | 12 | 13 |
|---|---|---|---|---|---|---|---|---|---|---|---|---|
| Hazaribagh | Pre-HYV | A.1 | 6.552 | -0.026 (0.144) | -0.368 (0.277) | 0.483*** (0.110) | - | 0.110 (0.114) | - | - | 0.918 | 14.29*** (4,5) |
| | | A.2 | 7.756 | -0.254 (0.143) | - | - | 0.271** (0.086) | 0.231 (0.137) | - | - | 0.828 | 9.689** (3,6)– |
| | | A.4 | 5.410 | - | - | - | - | 0.486** (0.194) | - | 0.030 (0.283) | 0.487 | 3.34* (2,7) |
| | HYV | A.1 | -1.496 | 0.483 (0.292) | 0.284 (0.298) | 0.505 (0.259) | - | 0.249 (0.172) | - | - | 0.925 | 15.78*** (4,5) |
| | | A.2 | 0.329 | 0.519 (0.268) | - | - | 0.409* (0.184) | 0.213 (0.150) | - | - | 0.922 | 23.64*** (3,6) |
| | | A.3 | 0.315 | 0.515 (0.298) | - | - | 0.391 (0.308) | 0.214 (0.165) | 0.036 (0.447) | - | 0.922 | 14.80*** (4,5) |
| | | A.4 | 1.600 | - | - | - | - | 0.212* (0.097) | - | 0.719*** (0.072) | 0.958 | 80.76*** (2,7) |
| | Combined | A.1 | 0.283 | 0.301* (0.145) | 0.151 (0.155) | 0.525*** (0.120) | - | 0.303** (0.111) | - | - | 0.857 | 22.65*** (4,15) |
| | | A.2 | 2.360 | 0.226 (0.143) | - | - | 0.368*** (0.067) | 0.359*** (0.110) | - | - | 0.835 | 27.09*** (3,160 |
| | | A.4 | 1.850 | - | - | - | - | 0.427*** (0.107) | - | 0.546*** (0.065) | 0.834 | 42.66*** (2,17) |

| 1 | 2 | 3 | 4 | 5 | 6 | 7 | 8 | 9 | 10 | 11 | 12 | 13 |
|---|---|---|---|---|---|---|---|---|---|---|---|---|
| Palamau | Pre-HYV | A.1 | 1.680 | 0.125<br>(0.094) | 0.312<br>(0.247) | 0.956***<br>(0.135) | - | 0.128<br>(0.243) | - | - | 0.968 | 38.61***<br>(4,5) |
| | | A.2 | 2.799 | 0.117<br>(0.119) | - | - | 0.787***<br>(0.137) | 0.176<br>(0.246) | - | - | 0.939 | 31.63***<br>(3,6) |
| | | A.4 | 2.926 | - | - | - | - | 0.921<br>(0.492) | - | 0.140<br>(0.359) | 0.370 | 2.058<br>(2,7) |
| | HYV | A.1 | 4.740 | 0.072<br>(0.357) | 0.562*<br>(0.245) | 0.156<br>(0.158) | - | 0.090<br>(0.153) | - | - | 0.757 | 3.898*<br>(4,5) |
| | | A.2 | 5.597 | 0.175<br>(0.375) | - | - | 0.274<br>(0.143) | 0.089<br>(0.164) | - | - | 0.664 | 3.962*<br>(3,6) |
| | | A.3 | 5.726 | 0.079<br>(0.390) | -<br>- | -<br>- | 0.153<br>(0.192) | 0.105<br>(0.166) | 0.312<br>(0.324) | - | 0.717 | 3.171<br>(4,5) |
| | | A.4 | 5.557 | - | - | - | - | 0.095<br>(0.170) | - | 0.409**<br>(0.160) | 0.507 | 3.604*<br>(2,7) |
| | Combined | A.1 | 2.876 | 0.137<br>(0.153) | 0.218<br>(0.141) | 0.483***<br>(0.119) | - | 0.215<br>(0.147) | - | - | 0.783 | 13.59***<br>(4,15) |
| | | A.2 | 4.799 | 0.087<br>(0.150) | - | - | 0.366***<br>(0.074) | 0.277*<br>(0.141) | - | - | 0.760 | 17.05***<br>(3,16) |
| | | A.4 | 5.176 | - | - | - | - | 0.332<br>(0.207) | - | 0.299***<br>(0.0821) | 0.454 | 7.104***<br>(2,17) |

| 1 | 2 | 3 | 4 | 5 | 6 | 7 | 8 | 9 | 10 | 11 | 12 | 13 |
|---|---|---|---|---|---|---|---|---|---|---|---|---|
| Ranchi | Pre-HYV | A.1 | 0.538 | -0.111<br>(0.144) | 0.103<br>(0.338) | 0.938***<br>(0.172) | - | 0.268<br>(0.271) | - | - | 0.908 | 12.56***<br>(4,5) |
| | | A.2 | 1.359 | -0.203<br>(0.164) | - | - | 0.703***<br>(0.147) | 0.642**<br>(0.232) | - | - | 0.841 | 10.60***<br>(3,6) |
| | | A.4 | 0.581 | - | - | - | - | 0.825*<br>(0.402) | - | 0.402<br>(0.230) | 0.461 | 3.000<br>(2,7) |
| | HYV | A.1 | -6.761 | 0.159<br>(0.227) | 1.013*<br>(0.421) | 1.001**<br>(0.293) | - | 0.446<br>(0.229) | - | - | 0.955 | 27.22***<br>(4,5) |
| | | A.2 | -2.105 | 0.158<br>(0.207) | - | - | 1.005***<br>(0.254) | 0.444*<br>(0.194) | - | - | 0.955 | 43.73***<br>(3,6) |
| | | A.3 | -3.073 | 0.003<br>(0.253) | - | - | 0.921**<br>(0.265) | 0.525**<br>(0.207) | 0.497<br>(0.471) | -<br>- | 0.964<br>0.964 | 33.68***<br>(4,5) |
| | | A.4 | 2.010 | - | - | - | - | 0.791*<br>(0.0394) | - | 0.286***<br>(0.078) | 0.701 | 8.228**<br>(2,7) |
| | Com-bined | A.1 | -5.338 | 0.117<br>(0.115) | 0.718***<br>(0.198) | 1.031***<br>(0.148) | - | 0.466**<br>(0.185) | - | - | 0.947 | 66.97***<br>(4,15) |
| | | A.2 | -1.627 | 0.085<br>(0.113) | - | - | 0.912***<br>(0.109) | 0.582***<br>(0.158) | - | - | 0.941 | 86.86***<br>(3,16) |
| | | A.4 | 0.945 | - | - | - | - | 0.789<br>(0.296) | - | 0.401***<br>(0.053) | 0.780 | 30.28***<br>(2,17) |

| 1 | 2 | 3 | 4 | 5 | 6 | 7 | 8 | 9 | 10 | 11 | 12 | 13 |
|---|---|---|---|---|---|---|---|---|---|---|---|---|
| Singhbhum | Pre-HYV | A.1 | -3.471 | 0.280 (0.261) | 0.815 (0.536) | 0.351 (0.310) | - | 0.452 (0.527) | - | - | 0.709 | 3.002 (4,5) |
| | | A.2 | 0.371 | 0.277 (0.247) | - | - | 0.501** (0.190) | 0.254 (0.402) | - | - | 0.687 | 4.401* (3,6) |
| | | A.4 | 1.973 | - | - | - | - | 0.627 (0.495) | - | 0.195* (0.089) | 0.445 | 2.808 (2,7) |
| | HYV | A.1 | -7.674 | 0.482* (0.189) | 0.598 (0.494) | 0.529* (0.233) | - | 0.950** (0.258) | - | - | 0.933 | 17.66*** (4,5) |
| | | A.2 | -4.867 | 0.543*** (0.136) | - | - | 0.459** (0.165) | 0.945*** (0.225) | - | - | 0.935 | 29.35*** (3,6) |
| | | A.3 | -4.882 | 0.539* (0.223) | - | - | 0.458* (0.190) | 0.945** (0.250) | 0.011 (0.507) | - | 0.936 | 18.35*** (4,5) |
| | | A.4 | 1.186 | - | - | - | - | 0.166 (0.111) | - | 0.772*** (0.041) | 0.980 | 189.8*** (2,7) |
| | Combined | A.1 | -6.795 | 0.437*** (0.127) | 0.999*** (0.223) | 0.361** (0.150) | - | 0.707 (0.194) | - | - | 0.935 | 55.55*** (4,5) |
| | | A.2 | -3.915 | 0.573*** (0.128) | - | - | 0.527*** (0.128) | 0.636*** (0.216) | - | - | 0.914 | 56.75*** (3,16) |
| | | A.4 | 2.353 | - | - | - | - | 0.403 (0.288) | - | 0.431*** (0.050) | 0.821 | 39.21*** (2,17) |

| 1 | 2 | 3 | 4 | 5 | 6 | 7 | 8 | 9 | 10 | 11 | 12 | 13 |
|---|---|---|---|---|---|---|---|---|---|---|---|---|
| Dhanbad | Pre-HYV | A.1 | 9.095 | 0.095 (0.492) | -1.451 (3.308) | -0.062 (0.890) | - | 0.244 (0.044) | - | - | 0.119 | 0.168 (4,5) |
| | | A.2 | 4.763 | 0.183 (0.391) | - | - | -0.313 (0.496) | 0.167 (0.560) | - | - | 0.097 | 0.217 (3,6) |
| | | A.4 | 0.472 | - | - | - | - | 0.144 (0.201) | - | 0.784*** (0.141) | 0.815 | 15.47*** (2,7) |
| | HYV | A.1 | -1.034 | 0.402 (0.225) | 0.300 (0.752) | 1.182** (0.428) | - | 0.833 (0.463) | - | - | 0.878 | 9.068** (4,5) |
| | | A.2 | -6.792 | 0.351 (0.222) | - | - | 0.985** (0.390) | 0.714 (0.454) | - | - | 0.850 | 11.48*** (3,6) |
| | | A.3 | -5.679 | 0.432 (0.282) | - | - | 1.142* (0.509) | 0.740 (0.487) | -0.590 (1.108) | - | 0.860 | 7.653** (4,5) |
| | | A.4 | 0.369 | - | - | - | - | -0.027 (0.053) | - | 0.975*** (0.018) | 0.998 | 1541.9*** (2,7) |
| | Combined | A.1 | -3.915 | 0.446* (0.225) | 0.782 (0.497) | 0.439 (0.434) | - | 0.150 (0.400) | - | - | 0.752 | 11.42*** (4,15) |
| | | A.2 | -1.422 | 0.465** (0.215) | - | - | 0.593* (0.283) | 0.144 (0.390) | - | - | 0.748 | 15.95*** (3,16) |
| | | A.4 | 0.237 | - | - | - | - | 0.108 (0.114) | - | 0.878*** (0.034) | 0.976 | 351.7*** (2,17) |

However, since the $R^2$ value in equation A.4 with irrigation variable exceeds the $R^2$ value estimated for equation A.1, A.2 and A.3 without the irrigation variable during the HYV period in the districts of Santhal Pargana, Hazaribagh, Singhbhum and Dhanbad, it can be taken to indicate that the irrigation factor plays quite a dominating role in the planting decision of the farmers in respect of the wheat crop during the HYV period. Farmers' decision for planting wheat is also influenced by the availability of rainfall during the sowing season as indicated by the positive (and in most cases significant) coefficient of the pre-sowing rainfall variable. It is quite interesting to note that the absolute value of the coefficient of pre-sowing rainfall variable increases in most cases when irrigation variable is included in the regression with the pre-sowing rainfall variable. This might indicate the complementary characteristics of the pre-sowing rainfall and the irrigation facilities i.e. the utilisation of pre-sowing rainfall for growing wheat is subject to the availability of irrigation facilities during the rabi season.

Fertiliser-price coefficients in equation A.3 estimated for the HYV period are positive in all the districts except in Dhanbad which exhibits a negative coefficient. However, all the coefficients are insignificant indicating the fact that the rise in fertiliser prices have not (adversely) affected the area under wheat. Table 5.6 showing the regression results for barley indicates that the farmers of Santhal Pargana, Hazaribagh and Palamau have traditionally planted the crop subject to the availability of rainfall during the sowing season. Except the lagged dependent variable and the pre-sowing rainfall variable, no other variables have statistically significant coefficients.

The overall homogeneity of regressions in the two sub-periods is accepted in the case of barley but rejected in the case of wheat for the districts of Hazaribagh and Palamau (details are given in Table 5.7). The overall homogeneity of the regressions between the districts for wheat and barley and for kharif cereals is, however, rejected.[4]

4. It should be noted that the validity of these tests, as pointed out earlier, is not fully guaranteed in a small sample situation because of the presence of the lagged endogenous variable in the regression-equation. As far as the second test that is the test for inter-district homogeneity of the regressions is concerned, the number of observations is twenty and hence not very small. In this case, however, there is an additional difficulty in the application of the test as Bartlett's modification of the likelihood ratio statistics [68 p.245], used for this purpose, indicates that the residual variances of the regression between the districts significantly differ. Bartlett's test is also, however, only asymptotically valid.

Thus we find that the price responsiveness of the Chotanagpur farmers is very much constrained by the subsistence characteristics of the crops and their dependence on monsoon. All price and gross return coefficients are insignificant except in a few cases. However,

TABLE 5.6 : **Showing acreage response equations estimated for individual crops in the two sub-periods and combined period - Barley**

| Constant, Independent Variables, $R^2$ and F | CHOTANAGPUR REGION AS A WHOLE (PANEL DATA) | | | | | |
|---|---|---|---|---|---|---|
| | Pre-HYV period | | HYV period | | Combined period | |
| | Eqn. No. A.1 | Eqn. No. A.2 | Eqn. No. A.1 | Eqn. No. A.2 | Eqn. No. A.1 | Eqn. No. A.2 |
| 1 | 2 | 3 | 4 | 5 | 6 | 7 |
| 1. Constant | -0.135 | -0.438 | 1.106 | 0.880 | 0.902 | 0.428 |
| 2. Log $A_{rt-1}$ | 0.880*** (0.097) | 0.873*** (0.095) | 0.947*** (0.044) | 0.948*** (0.043) | 0.900*** (0.052) | 0.899*** (0.051) |
| 3. Log $P_{rt-1}$ | 0.004 (0.373) | - | -0.019 (0.108) | - | -0.081 (0.080) | - |
| 4. Log $Y_{rt-1}$ | -0.253 (0.276) | - | -0.044 (0.155) | - | -0.106 (0.156) | - |
| 5. Log $G_{rt-1}$ | - | -0.159 (0.208) | - | -0.025 (0.093) | - | -0.086 (0.074) |
| 6. Log $R_{rt}$ | 0.473* (0.265) | 0.412* (0.235) | -0.056 (0.102) | -0.053 (0.098) | 0.166 (0.124) | 0.162 (0.120) |
| 7. $R^2$ | 0.769 | 0.762 | 0.953 | 0.953 | 0.850 | 0.850 |
| 8. F $(v_1,v_2)$ | 20.89*** (4,25) | 28.56*** (3,26) | 123.3*** (4,25) | 170.8*** (3,26) | 78.17*** (4,55) | 106.06*** (3,56) |
| | SANTHAL PARGANA DISTRICT | | | | | |
| 1. | 3.867 | 5.136 | 5.836 | 5.309 | 5.817 | 5.842 |
| 2. | 0.241 (0.231) | 0.241 (0.210) | 0.115 (0.708) | 0.232 (0.654) | 0.283 (0.167) | 0.295 (0.169) |
| 3. | 0.271 | - (0.440) | -0.124 | - (0.219) | -0.173 | - (0.106) |
| 4. | 0.293 | - (0.367) | 0.128 | - (0.319) | 0.158 | - (0.227) |
| 5. | - | 0.284 (0.261) | - - | -0.040 (0.172) | - | -0.100 (0.089) |

| 1 | 2 | 3 | 4 | 5 | 6 | 7 |
|---|---|---|---|---|---|---|
| 6. | 0.002<br>(0.332) | 0.011<br>(0.230) | 0.247<br>(0.255) | 0.250<br>(0.242) | 0.017<br>(0.190) | 0.138<br>(0.165) |
| 7. | 0.324 | 0.324 | 0.267 | 0.200 | 0.353 | 0.286 |
| 8. | 0.599<br>(4,5) | 0.960<br>(3,6) | 0.457<br>(4,5) | 0.499<br>(3,6) | 2.042<br>(4,15) | 2.138<br>(3,16) |
| | | | **HAZARIBAGH DISTRICT** | | | |
| 1. | 0.969 | 1.975 | 2.830 | 3.373 | 3.076 | 3.347 |
| 2. | 0.135<br>(0.374) | 0.177<br>(0.339) | 0.464<br>(0.372) | 0.438<br>(0.340) | 0.269<br>(0.189) | 0.318<br>(0.184) |
| 3. | 0.389<br>(0.654) | - | -0.032<br>(0.162) | - | -0.110<br>(0.105) | - |
| 4. | 0.010<br>(0.383) | - | 0.068<br>(0.203) | - | 0.121<br>(0.186) | - |
| 5. | - | 0.116<br>(0.287) | - | 0.006<br>(0.120) | - | -0.050<br>(0.089) |
| 6. | 0.748*<br>(0.359) | 0.669*<br>(0.297) | 0.178<br>(0.157) | 0.159<br>(0.138) | 0.403**<br>(0.145) | 0.392**<br>(0.145) |
| 7. | 0.531 | 0.510 | 0.549 | 0.534 | 0.528 | 0.495 |
| 8. | 1.414<br>(4,5) | 2.079<br>(3,6) | 1.518<br>(4,5) | 2.296<br>(3,6) | 4.190**<br>(4,15) | 5.199**<br>(3,16) |
| | | | **PALAMAU DISTRICT** | | | |
| 1. | 2.574 | 2.282 | 11.630 | 8.429 | 7.918 | 7.598 |
| 2. | 0.221<br>(0.398) | 0.216<br>(0.364) | 0.171<br>(0.479) | 0.324<br>(0.457) | 0.112<br>(0.261) | 0.109<br>(0.248) |
| 3. | 0.013<br>(0.695) | - | -0.069<br>(0.244) | - | -0.059<br>(0.155) | - |
| 4. | -0.129<br>(0.474) | - | -0.399<br>(0.315) | - | -0.074<br>(0.272) | - |
| 5. | - | -0.084<br>(0.370) | - | -0.171<br>(0.224) | - | -0.062<br>(0.139) |
| 6. | 0.994<br>(0.546) | 0.999*<br>(0.499) | -0.166<br>(0.208) | -0.168<br>(0.209) | 0.236<br>(0.226) | 0.235<br>(0.218) |

| 1 | 2 | 3 | 4 | 5 | 6 | 7 |
|---|---|---|---|---|---|---|
| 7. | 0.442 | 0.438 | 0.307 | 0.163 | 0.096 | 0.096 |
| 8. | 0.989 | 1.559 | 0.558 | 0.390 | 0.400 | 0.567 |
| | (4,5) | (3,6) | (4,5) | (3,6) | (4,15) | (3,16) |

the absolute values of these coefficients have increased in the HYV period as compared to those in the pre-HYV period in most cases but the increase is not very marked. We find lower price elasticities for major cereals (i.e. winter rice) than for minor cereals and higher price elasticities for rabi cereals particularly wheat than for kharif cereals. Pre-sowing rainfall appears to be an important factor influencing farmers' decisions for planting kharif crops whereas considerable increase in the area under wheat during the HYV period in Chotanagpur districts is mainly attributable to the increased irrigation facilities provided to the farmers during the rabi season. The overall homogeneity of regressions between the two sub-periods is accepted in all the cases except in a few cases. On the whole non-price factors appear to be more important than the price factor in influencing farmers' decisions for growing individual crops. However, there is a possibility that the price coefficients in the above regressions are not significant because as argued below the price responsiveness may be non-symmetric rather than symmetric. The next three sections look into this point.

## 5.3 NON-SYMMETRIC ACREAGE RESPONSE FOR SUBSISTENCE CROPS

### 5.3.1 **Introduction**

Supply response studies generally assume symmetric responses to price changes in the sense that output increases when price rises and declines when price falls. This may be so if the crops are grown mainly for the market but in the case of subsistence-crops, which are grown mainly for farmers' own consumption, farmers' behaviour may be different when price rises and when price falls. The farmers may be interested in increasing the output when price rises but they may not reduce the output even if price falls if the crops are mainly grown for their own consumption. We have examined this differential behaviour of farmers in terms of acreage allocation to different individual crops, when price rises and when price falls, in the next two sections.

### 5.3.2 Alternative Specifications of the Regression Equations

Two alternative equations are formulated for this purpose namely,

$$A.5 \ \text{Log}\frac{A_{rt}}{A_{rt-1}} = a + b_1\left(\text{Log}\frac{P_{rt-1}}{P_{rt-2}}\right)Z_1 + b_2\left(\text{Log}\frac{P_{rt-1}}{P_{rt-2}}\right)Z_2$$

$$+ b_3 \text{Log}\frac{R_{rt}}{R_{rt-1}} + b_4 \text{Log}\frac{I_t}{I_{t-1}} + b_5 \text{Log}\frac{A_{rt-1}}{A_{rt-2}} + u_t$$

or Simply

$$a_{rt} = a + b_1 e_1 + b_2 e_2 + b_3 r_{rt} + b_4 i_t + b_5 a_{rt-1} + u_t$$

$$A.6 \ \text{Log}\frac{A_{rt}}{A_{rt-1}} = a + b_1\left(\text{Log}\frac{G_{rt-1}}{G_{rt-2}}\right)Z'_1 + b_2\left(\text{Log}\frac{G_{rt-1}}{G_{rt-2}}\right)Z'_2$$

$$+ b_3 \text{Log}\frac{R_{rt}}{R_{rt-1}} + b_4 \text{Log}\frac{I_t}{I_{t-1}} + b_5 \text{Log}\frac{A_{rt-1}}{A_{rt-2}} + u_t$$

or Simply

$$a_{rt} = a + b_1 e'_1 + b_2 e'_2 + b_3 r_{rt} + b_4 i_t + b_5 a_{rt-1} + u_t$$

Where, $Z_1 = 1$ if $\frac{P_{rt-1}}{P_{rt-2}} > 1$ ; $Z_2 = 1$ if $\frac{P_{rt-1}}{P_{rt-2}} \leq 1$

= 0 if ohterwise = 0 if otherwise

and $Z'_1 = 1$ if $\frac{G_{rt-1}}{G_{rt-2}} > 1$ ; $Z'_2 = 1$ if $\frac{G_{rt-1}}{G_{rt-2}} \leq 1$

= 0 if otherwise = 0 if otherwise.

$A_{rt}$, $P_{rt}$, $G_{rt}$, $R_{rt}$, $I_t$ and $u_t$ have the same meaning as explained earlier. As is evident in equation A.5 and A.6

$$a_{rt} = \text{Log}\left(\frac{A_{rt}}{A_{rt-1}}\right);\ a_{rt-1} = \text{Log}\left(\frac{A_{rt-1}}{A_{rt-2}}\right)$$

$$r_{rt} = \text{Log}\left(\frac{R_{rt}}{R_{rt-1}}\right);\ i_t = \text{Log}\left(\frac{I_t}{I_{t-1}}\right)$$

$$e_1 = \left(\text{Log}\frac{P_{rt-1}}{P_{rt-2}}\right)Z_1;\ e_2 = \left(\text{Log}\frac{P_{rt-1}}{P_{rt-2}}\right)Z_2$$

$$e'_1 = \left(\text{Log}\frac{G_{rt-1}}{G_{rt-2}}\right)Z'_1;\ e'_2 = \left(\text{Log}\frac{G_{rt-1}}{G_{rt-2}}\right)Z'_2$$

Symmetric response, here, implies the equality of $b_1$ and $b_2$ (i.e. $b_1 = b_2$) and non-symmetric response, inequality of $b_1$ and $b_2$ (i.e. $b_1 \neq b_2$). We have tested the equality of $b_1$ and $b_2$ in equation A.5 and A.6

### 5.3.3 Regression Results

Regression equations A.5 and A.6 are estimated for two major cereals i.e. winter rice and wheat for each district and for Chotanagpur region pooling the districts' data, for the combined period. The irrigation variable is relevant for wheat only and is therefore used in the equation for wheat. Regression results are presented in Tables 5.8 and 5.9. $R^2$'s of these regressions (including that of Chotanagpur region as a whole with panel data) are not very high. It ranges between 0.387 to 0.738 in the case of winter rice and from 0.394 to 0.898 in the case of wheat. The F-ratios associated with the test for equality of two regression coefficients (i.e. $b_1 = b_2$) are computed and presented in Table 5.10.

TABLE 5.7: **Showing computed F ratios associated with the test of overall homogeneity of acreage response equations for individual crops between the two sub-periods (Equation No. A.1)**

| District/ Region | Computed F statistics with degrees of freedom in brackets for | | | | | |
|---|---|---|---|---|---|---|
| | Winter rice | Autumn rice | Maize | Ragi | Wheat | Barley |
| 1 | 2 | 3 | 4 | 5 | 6 | 7 |
| Santhal-Pargana | 1.051 (5,10) | 2.949* (5,10) | 1.425 (5,10) | 0.582 (5,10) | 1.761 (5,10) | 0.652 (5,10) |
| Hazaribagh | 4.636** (5,10) | 1.392 (5,10) | 1.158 (5,10) | 0.215 (5,10) | 3.025* (5,10) | 0.530 (5,10) |
| Palamau | 1.709 (5,10) | 1.542 (5,10) | 2.053 (5,10) | 1.410 (5,10) | 4.414** (5,10) | 1.279 (5,10) |
| Ranchi | 0.973 (5,10) | 1.475 (5,10) | 4.955** (5,10) | 0.234 (5,10) | 2.260 (5,10) | - |
| Singhbhum | 1.502 (5,10) | 0.185 (5,10) | 0.405 (5,10) | - | 0.565 (5,10) | - |
| Dhanbad | 2.036 (5,10) | - | 1.418 (5,10) | 0.824 (5,10) | 1.852 (5,10) | - |
| Chotanagpur region as a whole | 1.270 (5,110) | 0.101 (5,90) | 0.849 (5,110) | 0.299 (5,90) | 1.352 (5,110) | 0.939 (5,50) |

Note: $$F = \frac{(S_2 - S_1) / (NK-K)}{S_1 / (\sum_{i=1}^{N} T_i - NK)} \sim F_{(v_1, v_2)}$$

Where,

$S_2$ = Resideual sum of squares in the regression for the combined period

$S_1 = \sum_{i=i}^{N} RSS_i$ = sum of the residual sum of squares in the regression for the two sub-periods.

N = Number of periods,

K = Number of parameters including constant term and

$T_i$ = Number of observations in the ith period.

The equality of the two regression coefficients (i.e. $b_1$ and $b_2$) is rejected in the case of winter rice of Palamau and Ranchi and in the case of wheat for Hazaribagh, Dhanbad and Chotanagpur region as a whole (panel data). In other cases it is accepted. One would expect a positive sign for $b_1$ (the coefficient of $e_1$ and $e_2$) and significantly so if, in fact, the farmers are responsive to price changes and zero or negative sign for $b_2$ (the coefficient of $e_2$ and $e'_2$. The coefficients of $e_1$ are insignificant and in most cases negative indicating that price itself is not an important factor in explaining the changes in the area under the crop. The negative coefficients of $e_1$ may be due to the non-availability of sufficient rainfall-water during the sowing season. This is also borne out by the positive and in most cases significant coefficients of the pre-sowing rainfall variable. The coefficients of $e_2$ are also insignificant except in a few cases. It should be noted, however, that in most cases equation A.6 gives a better fit (considering $R^2$ values) than equation A.5. This is specially so for the wheat crop indicating that the farmers are responsive to the gross return rather than to price alone as noted in our earlier discussions also. The coefficients of $e'_1$ in equation A.6 estimated for the wheat crop (see Table 5.9) are positive in all the cases and significant in most of the cases.

It should be noted that in the case of winter rice, which is the major cereal, for Singhbhum district both the coefficients ($e_1$ and $e_2$ in equation A.5 and $e'_1$ and $e'_2$ in equation A.6) are negative and the coefficients of $e_2$ and $e'_2$ are significant at 5 per cent level of significance which might indicate the backward sloping supply behaviour of the farmers. Their behaviour appears to be perverse but quite consistent with the socio-economic attitudes of the tribal (Ho)[5] farmers concentrated in the district. Winter rice is the single major crop grown by the farmers for their own consumption and also for cash requirements which is primarily fixed due to their limited taste and natural conservatism.[6] When the price of winter rice rises, a small surplus would be sufficient for meeting the case requirements for purchasing other goods and the farmers might prefer leisure to the labour involved in cultivating more land to get a higher output. But if price falls they require additional output even for meeting the basic minimum of their requirements and hence increase the area under the crop for additional output.

---

5. Name of a tribe.

6. These attitudes are common for all the tribals but specially so for the Ho tribals concentrated in Singhbhum. This is because of their rigid cultural attitudes and isolation from the Hindu society resulting in a smaller social change among the Ho tribals than in other tribals of Chotanagpur as noted earlier in Chapter 2.

TABLE 5.8: **Showing non-symmetric acreage response equations estimated for the districts of Chotanagpur region in the combined period for Winter rice**

| District/ Region | Eqn. No. | Cons-tant | Coefficients of independent variables $e_1$ | $e_2$ | $e_1$ | $e_2$ | $r_{rt}$ | $a_{rt-1}$ | $R^2$ | F $(v_1,v_2)$ |
|---|---|---|---|---|---|---|---|---|---|---|
| 1 | 2 | 3 | 4 | 5 | 6 | 7 | 8 | 9 | 10 | 11 |
| Santhal-Pargana | A.5 | 0.005 | -0.061 (0.174) | 0.044 (0.156) | - | - | 0.196*** (0.066) | -0.556*** (0.162) | 0.578 | 5.124*** (4,15) |
| | A.6 | 0.009 | - | - | -0.070 (0.152) | 0.183 (0.207) | 0.192*** (0.066) | -0.520*** (0.175) | 0.594 | 5.302*** (4,15) |
| Hazaribagh | A.5 | -0.002 | -0.035 (0.078) | -0.033 (0.076) | - | - | 0.164*** (0.046) | -0.184 (0.207) | 0.588 | 5.349*** (4,15) |
| | A.6 | -0.005 | - | - | -0.015 (0.055) | -0.059 (0.054) | 0.167*** (0.046) | -0.174 (0.198) | 0.616 | 6.014*** (4,15) |
| Palamau | A.5 | 0.030 | -0.268 (0.143) | 0.199 (0.137) | - | - | 0.154** (0.065) | -0.220 (0.159) | 0.506 | 3.836** (4,15) |
| | A.6 | 0.008 | - | - | 0.077 (0.045) | -0.036 (0.072) | 0.135* (0.070) | -0.179 (0.162) | 0.479 | 3.452** (4,15) |
| Ranchi | A.5 | 0.007 | -0.024 (0.042) | 0.128** (0.048) | - | - | 0.035 (0.022) | -0.606*** (0.130) | 0.738 | 10.57*** (4,15) |
| | A.6 | 0.002 | - | - | 0.016 (0.078) | 0.045 (0.044) | 0.036 (0.029) | -0.653*** (0.146) | 0.651 | 6.992*** (4,15) |

| District/ Region | Eqn. No. | Cons-tant | Coefficients of independent variables | | | | | | $R^2$ | F $(v_1,v_2)$ |
|---|---|---|---|---|---|---|---|---|---|---|
| | | | $e_1$ | $e_2$ | $e_1$ | $e_2$ | $r_{rt}$ | $a_{rt-1}$ | | |
| 1 | 2 | 3 | 4 | 5 | 6 | 7 | 8 | 9 | 10 | 11 |
| Singhbhum | A.5 | -0.002 | -0.015 (0.071) | -0.187** (0.069) | - | - | -0.050 (0.037) | -0.223 (0.148) | 0.626 | 5.854*** (4,14) |
| | A.6 | 0.001 | - | - | -0.071 (0.069) | -0.129** (0.059) | -0.015 (0.064) | -0.533** (0.222) | 0.581 | 4.859*** (4,14) |
| Dhanbad | A.5 | -0.011 | 0.003 (0.172) | 0.117 (0.266) | - | - | 0.180* (0.088) | -0.404 (0.317) | 0.482 | 3.253** (4,14) |
| | A.6 | -0.017 | - | - | 0.026 (0.099) | 0.015 (0.162) | 0.189** (0.082) | -0.464 (0.294) | 0.478 | 3.196** (4,14) |
| Chotanagpur region as a whole (pooled) | A.5 | 0.004 | -0.064 (0.051) | 0.019 (0.052) | - | - | 0.137*** (0.024) | -0.408*** (0.077) | 0.387 | 17.87*** (4,113) |
| | A.6 | -0.005 | - | - | 0.033 (0.026) | -0.041 (0.036) | 0.129*** (0.025) | -0.411*** (0.076) | 0.389 | 18.04*** (4,113) |

Note: Number of observations (n) are 20 for each district except for Singhbhum and Dhanbad for which n = 19. There was territorial re-adjustment of area in 1956 between Bihar and West Bengal which affected the area of Singhbhum and Dhanbad with the result that for the 1954-55 data for the two districts with area as adjusted in 1956 are not available.

## 5.4 OVERALL ACREAGE RESPONSE OF TOTAL CEREALS

### 5.4.1. Introduction

The analysis of aggregate agricultural supply response both in terms of acreage and in terms of output to agricultural prices is, of course, the prime consideration if one is concerned with increasing the aggregate level of agricultural production through price policy. Mellor [82] has noted that we are in need of careful studies on aggregate supply response for the agricultural sector which can give us a more accurate idea as to the extent to which price changes do in-fact affect aggregate agricultural production. He further speculates that the aggregate supply response to price is lower in low-income traditional agriculture than in high-income modern agriculture.

### 5.4.2 Alternative Specifications of the Regression Equations

The following are the three alternative equations estimated for this purpose

$$\text{B.1 } \operatorname{Log} A_t = a + b_1 \operatorname{Log} A_{t-1} + b_2 \operatorname{Log} P_{t-1} + b_3 \operatorname{Log} Y_{t-1} + b_4 \operatorname{Log} R_t + u_t$$

$$\text{B.2 } \operatorname{Log} A_t = a + b_1 \operatorname{Log} A_{t-1} + b_2 \operatorname{Log} P^R_{t-1} + b_3 \operatorname{Log} Y_{t-1} + b_4 \operatorname{Log} R_t + u_t$$

$$\text{B.3 } \operatorname{Log} A_t = a + b_1 \operatorname{Log} A_{t-1} + b_2 \operatorname{Log} G_{t-1} + b_3 \operatorname{Log} R_t + u_t$$

Where $A_t$ indicates aggregate area under total cereals and $P_t$, $Y_t$, and $G_t$ are respectively average price, average yield rate and average gross return of total cereals, $R_t$ is the pre-sowing rainfall relevant to the total cereals and $P^R_t$ is relative price of total cereals (i.e. average price of total cereals relative to the Economic Advisor's All Commodity Wholesale Price Index) and $u_t$ is the error term.

TABLE 5.9: **Showing non-symmetric acreage response equations estimated for the districts of Chotanagpur region in the combined period for Wheat**

| District/ Region | Eqn. No. | Cons- tant | Coefficients of independent variables | | | | | | | $R^2$ | F $(v_1,v_2)$ |
|---|---|---|---|---|---|---|---|---|---|---|---|
| | | | $e_1$ | $e_2$ | $e_1$ | $e_2$ | $i_t$ | $r_{rt}$ | $a_{rt-1}$ | | |
| 1 | 2 | 3 | 4 | 5 | 6 | 7 | 8 | 9 | 10 | 11 | 12 |
| Santhal-Pargana | A.5 | 0.030 | 0.015<br>(0.038) | -0.532<br>(0.741) | - | - | 0.256**<br>(0.104) | 0.301**<br>(0.122) | -0.329*<br>(0.174) | 0.646 | 5.112***<br>(5,14) |
| | A.6 | -0.001 | - | - | 0.410**<br>(0.165) | -0.104<br>(0.208) | 0.232***<br>(0.094) | 0.281**<br>(0.110) | -0.239<br>(0.148) | 0.731 | 7.596***<br>(5,14) |
| Hazaribagh | A.5 | -0.392 | 0.190<br>(0.346) | -1.799**<br>(0.711) | - | - | 0.278***<br>(0.091) | 0.247***<br>(0.074) | -0.633***<br>(0.124) | 0.810 | 11.99***<br>(5,14) |
| | A.6 | 0.026 | - | - | 0.185<br>(0.170) | 0.379*<br>(0.179) | 0.226**<br>(0.097) | 0.171**<br>(0.076) | -0.318*<br>(0.177) | 0.794 | 10.75***<br>(5,14) |
| Palamau | A.5 | 0.295 | -0.133<br>(0.629) | -0.106<br>(1.575) | - | - | 0.262<br>(0.218) | 0.318**<br>(0.127) | -0.570**<br>(0.217) | 0.565 | 3.641**<br>(5,14) |
| | A.6 | 0.003 | - | - | 0.424**<br>(0.159) | 0.417*<br>(0.220) | -0.038<br>(0.186) | 0.259**<br>(0.088) | -0.172<br>(0.186) | 0.784 | 10.177***<br>(5,14) |
| Ranchi | A.5 | 2.296 | -1.093<br>(0.949) | -0.189<br>(1.442) | - | - | 0.251<br>(0.324) | 0.138<br>(0.250) | -0.394<br>(0.233) | 0.476 | 2.543*<br>(5,14) |
| | A.6 | -0.001 | - | - | 0.716**<br>(0.252) | 1.060***<br>(0.304) | 0.266<br>(0.197) | 0.460***<br>(0.113) | -0.134<br>(0.141) | 0.814 | 12.162***<br>(5,14) |

| District/ Region | Eqn. No. | Cons- tant | Coefficients of independent variables $e_1$ | $e_2$ | $e_1$ | $e_2$ | $i_t$ | $r_{rt}$ | $a_{rt-1}$ | $R^2$ | F $(v_1,v_2)$ |
|---|---|---|---|---|---|---|---|---|---|---|---|
| 1 | 2 | 3 | 4 | 5 | 6 | 7 | 8 | 9 | 10 | 11 | 12 |
| Singhbhum | A.5 | 2.114 | -1.006 (0.766) | 1.827 (1.244) | - | - | 0.281 (0.172) | 0.143 (0.209) | 0.043 (0.234) | 0.457 | 2.192 (5,13) |
| | A.6 | 0.028 | - | - | 0.095 (0.245) | 0.205 (0.263) | 0.260 (0.194) | 0.290 (0.228) | 0.018 (0.247) | 0.394 | 1.694 (5,13) |
| Dhanbad | A.5 | 0.319 | -0.152 (0.455) | -0.370 (0.904) | - | - | 0.847*** (0.111) | -0.048 (0.167) | 0.078 (0.117) | 0.854 | 15.19*** (5,13) |
| | A.6 | -0.040 | - | - | 0.252 (0.178) | -0.478** (0.199) | 0.824*** (0.089) | -0.041 (0.123) | 0.031 (0.099) | 0.898 | 22.95*** (5,13) |
| Chotanagpur region as a whole (pooled) | A.5 | 0.003 | 0.001 (0.019) | -0.899*** (0.355) | - | - | 0.441*** (0.061) | 0.245*** (0.061) | -0.245*** (0.069) | 0.496 | 21.95*** (5,112) |
| | A.6 | -0.012 | - | - | 0.360*** (0.093) | 0.034 (0.112) | 0.396*** (0.058) | 0.228*** (0.059) | -0.172** (0.068) | 0.537 | 25.98*** (5,112) |

Note : Number of observations (n) are 20 for each district except for Singhbhum and Dhanbad for which n = 19. Same explanation applies here also as given in the note below the previous table.

### 5.4.3 Regression Results

Regression equations B.1 to B.3 are estimated for each district and for the Chotanagpur region (panel data) for the two sub-periods and for the combined period. Results are given in Table 5.11. $R^2$'s in

TABLE 5.10: **Showing computed F-ratios associated with the test for equality of the regression coefficients of $e_1$ and $e_2$ in Eqn. A.5 and $e'_1$ and $e'_2$ in Eqn. A.6 estimated for winter rice and wheat**

| District/ Region | Computed F statistics with degrees of freedom in brackets for | | | |
|---|---|---|---|---|
| | Winter rice | | Wheat | |
| | Eqn. No. A.5 | Eqn. No. A.6 | Eqn. No. A.5 | Eqn. No. A.6 |
| Santhal-Pargana | 0.144 (1,15) | 0.694 (1,15) | 1.082 (1,14) | 3.085 (1,14) |
| Hazaribagh | 0.0 (1,15) | 1.322 (1,15) | 6.364** (1,14) | 0.610 (1,14) |
| Palamau | 3.904* (1,15) | 1.240 (1,15) | 0.001 (1,14) | 0.0 (1,14) |
| Ranchi | 3.535* (1,15) | 0.207 (1,15) | 0.186 (1,14) | 0.700 (1,14) |
| Singhbhum | 2.146 (1,14) | 0.338 (1,14) | 2.650 (1,13) | 0.088 (1,13) |
| Dhanbad | 0.096 (1,14) | 0.002 (1,14) | 0.035 (1,13) | 5.859** (1,13) |
| Chotanagpur region as a whole | 0.929 (1,113) | 1.926 (1,113) | 1.939 (1,112) | 4.032** (1,112) |

Note : $F = \dfrac{(RRSS-URSS) / r}{URSS / (N - K)}$.

Where, RRSS = Restricted residual sum of squares
URSS = Un-restricted residual sum of squares
r = Number of restriction (r = 1 in this case)
N = Number of observations.
K = Number of parameters (in the original un-restricted equation) including constant term.

TABLE 5.11: **Showing overall acreage response equations estimated for the two sub- periods and combined period in the Chotanagpur region**

| Constant, Independen variables, $R^2$ and F | CHOTANAGPUR REGION AS A WHOLE (PANEL DATA) | | | | | | | | |
|---|---|---|---|---|---|---|---|---|---|
| | Pre-HYV period | | | HYV period | | | Combined period | | |
| | Eqn. No. B.1 | Eqn. No. B.2 | Eqn. No. B.3 | Eqn. No. B.1 | Eqn. No. B.2 | Eqn. No. B.3 | Eqn. No. B.1 | Eqn. No. B.2 | Eqn. No. B.3 |
| 1 | 2 | 3 | 4 | 5 | 6 | 7 | 8 | 9 | 10 |
| 1. Constant | 0.234 | 0.189 | -0.107 | -1.651 | -1.791 | -1.660 | -0.651 | -0.819 | -0.807 |
| 2. Log $A_{t-1}$ | 0.983*** (0.014) | 0.983*** (0.014) | 0.980*** (0.015) | 1.009*** (0.018) | 1.006*** (0.018) | 1.010*** (0.018) | 0.994*** (0.012) | 0.993*** (0.011) | 0.994*** (0.012) |
| 3. Log $P_{t-1}$ | 0.004 (0.046) | - | - | 0.014 (0.042) | - - | - - | -0.007 (0.018) | - | - |
| 4. Log $P^R_{t-1}$ | - | 0.021 (0.034) | - | - | 0.073 (0.065) | - | - | 0.044 (0.033) | - |
| 5. Log $Y_{t-1}$ | -0.035** (0.013) | -0.035** (0.013) | - | -0.005 (0.019) | -0.002 (0.018) | - – | -0.017 (0.011) | -0.016 (0.011) | - – |
| 6. Log $G_{t-1}$ | - | - | 0.001 (0.043) | - | - | 0.006 (0.018) | | | -0.002 (0.012) |
| 7. Log $R_t$ | 0.029 (0.051) | 0.027 (0.051) | 0.048 (0.054) | 0.212*** (0.067) | 0.206*** (0.066) | 0.208*** (0.066) | 0.122*** (0.042) | 0.121*** (0.042) | (0.126)*** (0.042) |
| 8. $R^2$ | 0.989 | 0.988 | 0.987 | 0.981 | 0.982 | 0.981 | 0.984 | 0.984 | 0.983 |
| 9. F $(v_1,v_2)$ | 1207.5*** (4,55) | 1216.2*** (4,55) | 1456.6*** (3,56) | 732.7*** (4,55) | 748.3*** (4,55) | 993.8*** (3,56) | 1736.5*** (4,115) | 1760.9*** (4,115) | 2290.5*** (3,116) |

SANTHAL PARGANA DISTRICT

| 1 | 2 | 3 | 4 | 5 | 6 | 7 | 8 | 9 | 10 |
|---|---|---|---|---|---|---|---|---|---|
| 1. | 16.651 | 16.470 | 17.162 | 17.208 | 14.054 | 17.525 | 13.891 | 13.927 | 13.926 |
| 2. | -0.331 (0.363) | -0.313 (0.360) | -0.396 (0.292) | -0.560 (0.438) | -0.222 (0.432) | -0.534 (0.390) | -0.150 (0.227) | -0.150 (0.227) | -0.154 (0.215) |
| 3. | 0.008 (0.070) | - | - | 0.233 (0.160) | - | - | 0.004 (0.044) | - | - |
| 4. | - | 0.015 (0.054) | - | - | -0.121 (0.270) | - | - | -0.006 (0.080) | - |
| 5. | -0.006 (0.012) | -0.007 (0.012) | - | 0.204 (0.317) | 0.161 (0.372) | - | -0.002 (0.019) | -0.002 (0.019) | - |
| 6. | - | - | 0.029 (0.005) | - | - | 0.198 (0.124) | - | - | 0.002 (0.030) |
| 7. | 0.127 (0.079) | 0.119 (0.086) | 0.144* (0.005) | 0.126 (0.163) | 0.187 (0.190) | 0.145 (0.144) | 0.175* (0.087) | 0.177* (0.092) | 0.177** (0.083) |
| 8. | 0.561 | 0.567 | 0.561 | 0.460 | 0.262 | 0.446 | 0.255 | 0.255 | 0.254 |
| 9. | 1.603 (4,5) | 1.41 (4,5) | 2.556 (3,6) | 1.065 (4,5) | 0.443 (4,5) | 1.613 (3,6) | 1.283 (4,15) | 1.282 (4,15) | 1.819 (3,16) |

HAZARIBAGH DISTRICT

| 1 | 2 | 3 | 4 | 5 | 6 | 7 | 8 | 9 | 10 |
|---|---|---|---|---|---|---|---|---|---|
| 1. | 13.119 | 13.434 | 13.136 | 12.474 | 12.477 | 8.311 | 10.323 | 8.662 | 8.526 |
| 2. | -0.086 (0.209) | -0.111 (0.200) | -0.114 (0.200) | -0.099 (0.709) | -0.109 (0.656) | 0.272 (0.417) | 0.113 (0.195) | 0.189 (0.215) | 0.265 (0.221) |
| 3. | -0.051 (0.078) | - | - | 0.032 (0.076) | - | - | -0.068* (0.032) | - | - |
| 4. | - | -0.048 (0.053) | - | - | 0.081 (0.106) | - | - | 0.037 (0.058) | - |
| 5. | 0.005 (0.076) | -0.004 (0.075) | - | 0.092 (0.298) | 0.091 (0.260) | - | 0.067 (0.076) | 0.093 (0.087) | - |
| 6. | - | - | -0.006 (0.067) | - | - | 0.010 (0.016) | - | - | 0.001 (0.017) |
| 7. | 0.125 (0.106) | 0.133 (0.094) | 0.160 (0.090) | 0.098 (0.111) | 0.097 (0.107) | 0.116 (0.100) | 0.108 (0.073) | 0.125 (0.083) | 0.113 (0.083) |
| 8. | 0.421 | 0.458 | 0.353 | 0.259 | 0.314 | 0.228 | 0.417 | 0.263 | 0.198 |
| 9. | 0.909 (4,5) | 1.059 (4,5) | 1.093 (3,6) | 0.439 (4,5) | 0.593 (4,5) | 0.589 (3,6) | 2.686* (4,15) | 1.341 (4,15) | 1.314 (3,16) |

PALAMAU DISTRICT

| 1 | 2 | 3 | 4 | 5 | 6 | 7 | 8 | 9 | 10 |
|---|---|---|---|---|---|---|---|---|---|
| 1. | 3.639 | 2.149 | 4.409 | 10.685 | 3.739 | 9.995 | 4.951 | 0.871 | 5.206 |
| 2 | 0.722***<br>(0.094) | 0.792***<br>(0.099) | 0.670***<br>(0.146) | -0.028<br>(0.327) | 0.448<br>(0.240) | 0.075<br>(0.373) | 0.535***<br>(0.082) | 0.744***<br>(0.082) | 0.538***<br>(0.092) |
| 3. | 0.036<br>(0.160 | - | - | 0.103<br>(0.054) | - | - | 0.037<br>(0.026) | - | - |
| 4. | - | 0.119<br>(0.118) | - | - | 0.271**<br>(0.104) | - | - | 0.250***<br>(0.083) | - |
| 5. | -0.152<br>(0.076) | -0.136*<br>(0.059) | - | 0.015<br>(0.009) | 0.030**<br>(0.010) | - | 0.009<br>(0.009) | 0.021**<br>(0.009) | - |
| 6. | - | - | -0.101<br>(0.132) | - | - | 0.077<br>(0.045) | - | - | 0.020<br>(0.022) |
| 7. | 0.086<br>(0.105) | 0.122<br>(0.088) | 0.043<br>(0.109) | 0.177<br>(0.118) | 0.260*<br>(0.116) | 0.085<br>(0.119) | 0.065<br>(0.075) | 0.176**<br>(0.077) | 0.033<br>(0.074) |
| 8. | 0.943 | 0.953 | 0.863 | 0.661 | 0.752 | 0.477 | 0.815 | 0.871 | 0.792 |
| 9. | 20.77***<br>(4,5) | 24.95***<br>(4,5) | 12.67***<br>(3,6) | 2.445<br>(4,5) | 3.789*<br>(4,5) | 1.826<br>(3,6) | 16.63***<br>(4,15) | 25.08***<br>(4,15) | 20.34***<br>(3,16) |

RANCHI DISTRICT

| 1 | 2 | 3 | 4 | 5 | 6 | 7 | 8 | 9 | 10 |
|---|---|---|---|---|---|---|---|---|---|
| 1. | 13.304 | 12.189 | 19.466 | 11.471 | 7.662 | 12.254 | 10.725 | 1.813 | 14.348 |
| 2. | -0.059 (0.640) | 0.041 (0.633) | -0.596* (0.295) | 0.045 (0.217) | 0.331 (0.389) | -0.020 (0.250) | 0.111 (0.197) | 0.819*** (0.196) | -0.178 (0.174) |
| 3. | 0.030 (0.053) | - | - | 0.098** (0.038) | - | - | 0.086*** (0.019) | - | - |
| 4. | - | 0.007 (0.041) | - | - | 0.009 (0.046) | - | - | 0.038 (0.029) | - |
| 5. | 0.028 (0.085) | 0.007 (0.083) | - | 0.039 (0.089) | 0.182 (0.110) | - | 0.038 (0.040) | 0.014 (0.057) | - |
| 6. | - | - | 0.131** (0.037) | - | - | 0.081** (0.024) | - | - | 0.087*** (0.013) |
| 7. | 0.051 (0.068) | 0.047 (0.070) | 0.083* (0.037) | 0.064 (0.064) | -0.006 (0.093) | 0.081 (0.037) | 0.055 (0.039) | 0.047 (0.057) | 0.073** (0.031) |
| 8. | 0.151 | 0.102 | 0.706 | 0.821 | 0.585 | 0.753 | 0.819 | 0.621 | 0.882 |
| 9. | 0.223 (4,5) | 0.142 (4,5) | 4.808** (3,6) | 5.778** (4,5) | 1.766 (4,5) | 6.126** (3,6) | 17.02*** (4,15) | 6.145*** (4,15) | 39.77*** (3,16) |

SINGHBHUM DISTRICT

| 1 | 2 | 3 | 4 | 5 | 6 | 7 | 8 | 9 | 10 |
|---|---|---|---|---|---|---|---|---|---|
| 1. | 18.436 | 18.882 | 9.381 | 15.236 | 17.006 | 14.148 | 11.443 | 16.156 | 12.382 |
| 2. | -0.405<br>(0.757) | -0.451<br>(0.571) | 0.276<br>(0.366) | -0.139<br>(0.444) | -0.261<br>(0.403) | -0.051<br>(0.424) | 0.136<br>(0.233) | -0.203<br>(0.251) | 0.063<br>(0.233) |
| 3. | -0.120<br>(0.137) | - | - | 0.038<br>(0.048) | - | - | 0.017<br>(0.020) | - | - |
| 4. | - | -0.115<br>(0.082 | - | - | -0.129<br>(0.085) | - | - | -0.113**<br>(0.045) | - |
| 5. | -0.006<br>(0.075) | -0.038<br>(0.073) | - | -0.065<br>(0.125) | -0.042<br>(0.102) | - | 0.034<br>(0.041) | -0.041<br>(0.045) | - |
| 6. | - | - | 0.078<br>(0.048) | - | - | -0.002<br>(0.030) | - | - | 0.016<br>(0.013) |
| 7. | 0.007<br>(0.146) | 0.052<br>(0.113) | -0.095<br>(0.079) | -0.051<br>(0.121) | -0.013<br>(0.111) | -0.093<br>(0.101) | -0.097*<br>(0.054) | -0.007<br>(0.061) | -0.072<br>(0.054) |
| 8. | 0.338 | 0.452 | 0.401 | 0.230 | 0.406 | 0.127 | 0.256 | 0.444 | 0.266 |
| 9. | 0.639<br>(4,5) | 1.031<br>(4,5) | 1.336<br>(3,6) | 0.374<br>(4,5) | 0.854<br>(4,5) | 0.293<br>(3,6) | 1.289<br>(4,15) | 2.996*<br>(4,15) | 1.931<br>(3,16) |

DHANBAD DISTRICT

| 1 | 2 | 3 | 4 | 5 | 6 | 7 | 8 | 9 | 10 |
|---|---|---|---|---|---|---|---|---|---|
| 1. | 8.175 | 8.497 | 6.196 | 9.896 | 8.164 | 10.385 | 7.145 | 1.416 | 7.727 |
| 2. | 0.079<br>(0.549) | 0.061<br>(0.542) | 0.327<br>(0.344) | -0.079<br>(0.660) | 0.073<br>(0.597) | -0.084<br>(0.502) | 0.230<br>(0.410) | 0.633*<br>(0.348) | 0.237<br>(0.282) |
| 3. | 0.038<br>(0.126) | - | - | -0.031<br>(0.107) | - | - | -0.126*<br>(0.066) | - | - |
| 4. | - | 0.027<br>(0.093) | - | - | 0.198<br>(0.231) | | | 0.141<br>(0.39) | - |
| 5. | 0.120<br>(0.159) | 0.118<br>(0.158) | - | -0.008<br>(0.274) | -0.127<br>(0.263) | | 0.047<br>(0.142) | 0.039<br>(0.149) | - |
| 6. | - | - | 0.022<br>(0.092) | - | - | -0.062<br>(0.072) | - | - | -0.111***<br>(0.037) |
| 7. | 0.193<br>(0.105) | 0.185<br>(0.102) | 0.189<br>(0.108) | 0.348<br>(0.194) | 0.350<br>(0.176) | 0.330*<br>(0.161) | 0.248**<br>(0.111) | 0.277**<br>(0.115) | 0.249**<br>(0.094) |
| 8. | 0.461 | 0.460 | 0.406 | 0.504 | 0.569 | 0.555 | 0.476 | 0.423 | 0.582 |
| 9. | 1.069<br>(4,5) | 1.067<br>(4,5) | 1.363<br>(3,6) | 1.319<br>(4,5) | 1.647<br>(4,5) | 2.496<br>(3,6) | 3.412**<br>(4,15) | 2.741*<br>(4,15) | 7.438**<br>(3,16) |

the regressions for the Chotanagpur region as a whole are very high but in the case of other regressions for the individual districts they are very low except in the case of the Palamau and Ranchi districts. The test of overall homogeneity of regressions in the two sub-periods is carried out for this case also and the results appear in Table 5.12.

TABLE 5.12: **Showing computed F-ratios associated with the test for overall homogeneity of overall acreage response equations in the two sub-periods (Eqn. No. B.1)**

| District/ Region | Computed F statistics with degrees of freedom in the brackets i.e. $F_{(v_1,v_2)}$ |
|---|---|
| Santhal-Pargana | 0.926 (5,10) |
| Hazaribagh | 1.405 (5,10) |
| Palamau | 2.210 (5,10) |
| Ranchi | 0.448 (5,10) |
| Singhbhum | 0.446 (5,10) |
| Dhanbad | 1.340 (5,10) |
| Chotanagpur region as a whole | 1.779 (5,110) |

The regression results estimated from the panel data for all the districts in the Chotanagpur region sufficiently indicate the dominance of the weather factor and the lagged dependent variable in the overall acreage response of the farmers. We find insignificant price and gross return coefficients in all the cases except in the case of Ranchi and Dhanbad districts. Ranchi shows significantly positive price and gross return coefficients in the HYV period and in the combined period equation whereas Dhanbad shows significantly negative price and gross return coefficients in the combined period equation. The relative price coefficient is significantly positive in Palamau district but significantly negative in Singhbhum in the regression for the combined period indicating the backward slopping supply behaviour. It has been noted earlier that the relative price of agricultural output has significantly declined in Singhbhum district in both the periods (see item 9 in Table 3.8). As a result of the decline in relative prices they require additional output even to maintain the minimum level of the present requirements of other (non-agricultural) goods. In the absence of other sources of income the farmers might have increased the area under total cereals for higher level of output.

The test of overall homogeneity of regressions in the two sub-periods is accepted in all the cases indicating that the overall acreage response pattern remains almost the same in the two sub-periods (details are given in Table 5.12).

## 5.5 CONCLUDING REMARKS

Regression results presented in this chapter reveal that the price responsiveness of Chotanagpur farmers specially of kharif crops like winter rice, autumn rice, maize and ragi is very much constrained by the subsistence characteristics of the crops and their dependence on monsoon rainfall. The subsistence farmers seem to respond to gross return of the crops rather than to the price factor alone. Increased irrigation facilities during the rabi season appears to be the most important factor influencing farmers' decision for planting wheat spetially during the HYV period. We find that the minor cereals exhibit higher price elasticity than major cereals and rabi cereals specially wheat exhibits a higher price response than kharif cereals. Price elasticity in the HYV period is higher than that in the pre-HYV period in most cases but the difference is not very marked.

Acreage response is found to be non-symmetric in the case of Palamau and Ranchi district for winter rice and in the case of the Hazaribagh, Dhanbad and Chotanagpur region as a whole for wheat. The analysis of the overall acreage response pattern is almost the same as that of winter rice because of its dominance in the total area sown. Backward slopping supply behaviour observed in the case of the Singhbhum district is quite consistent with the socio-economic attitudes of the tribal farmers concentrated in the district. On the whole we find that the non-price considerations, like availability of rainfall water irrigation facilities and traditional considerations are more important than price considerations in the decisions of the farmers. The response pattern remains almost the same in the two sub-periods. It has been noted that winter rice is grown singlely and has no substitute crop during the season. But other kharif cereals and rabi crops can be grown on the same plot of land and in the same season. For these crops the response may be in terms of relative acreage changes rather than in terms of absolute changes in acreages. To what extent the crop-substitutions are influenced by price and non-price factors is examined in the next chapter.

SIX

# CROPPING PATTERN (PAIR-WISE) RESPONSE IN CHOTANAGPUR REGION- Inter District and Inter Temporal Variations

## 6.1 INTRODUCTION

The analysis of cropping pattern in relation to price changes deserves adequate attention in the study of dynamic supply responses. Despite this fact many of the previous studies in the Indian context have neglected this aspect and concentrated only on the study of acreage response to one crop or a group of crops, which is but only one facet of supply responses. Acreage response equations, reflecting the effect of changes in price and non-price factors on the year-to-year variations in the area under one or a group of more than one crops, do not, however, bring out the changes in relative acreages under different crops in relation to changes in the relative prices. Infact, if total land is limited the acreage response to changes in relative prices may manifest itself in changes in the relative acreages only, whereas if additional land is available, it may manifest itself both in the acreage under different crops and changes in the relative acreages under the different crops. Additional land (and total land in the earlier sentence) is here taken to cover increase in cropping intensity also.

In fact the changes in relative acreages under different crops form an important aspect of the price-induced supply responses. If the farmers are responsive to prices, as would be the case if they are maximising profit (unit cost of production being assumed to be the same), one would expect then to substitute those crop(s) for which they find relatively higher prices or net returns, as compared to other substitutable crop(s) available to them. All these cause a change in the proportion of area under the crops and thus the cropping pattern might change over the period. These changes in the cropping pattern may be studied in two ways : Firstly, by intercrop substitution between pairs of substitutable crops, and secondly, in terms of the overall cropping pattern changes.

In the subsequent sections of this chapter we examine the cropping pattern response of Chotanagpur farmers to price and non-price factors at district level, in terms of crop-substitution between pairs of substitutable crops. The overall cropping pattern response to the price pattern (and also other factors determining the overall cropping pattern) is discussed in the next chapter.

TABLE 6.1: **Showing the group of substitutable crops during kharif and rabi seasons in the districts of Chotanagpur region**

| District | Crops included in the acreage substitution analyses during | |
|---|---|---|
| | Kharif season | Rabi season |
| Santhal Pargana | Autumn rice | Wheat |
| | Maize | Barley |
| | Ragi | Gram |
| Hazaribagh | Autumn rice | Wheat |
| | Maize | Barley |
| | Ragi | Gram |
| Palamau | Autumn rice | Wheat |
| | Maize | Barley |
| | Ragi | Gram |
| Ranchi | Autumn rice | Wheat |
| | Maize | Gram |
| | Ragi | .. |
| Singhbhum | Autumn rice | Wheat |
| | Maize | Gram |
| Dhanbad | Maize | .. |
| | Ragi | .. |

Notes : (i) Substitutable crops included in the group are the crops which can be grown on the same plot of land and in the same season.

(ii) Winter rice, the major kharif crop, could not be included in the group because the agro-climatic conditions required for growing the crop (Winter rice) preclude the possibilities of growing other kharif crops on the same plot of land.

(iii) Grouping of substitutable crops are made on the basis of informa tion given in Fertilizer Statistics 1969-70, p. 504.

## 6.2 SOME PRELIMINARY CONSIDERATIONS RELATED TO CROP-SUBSTITUTION

As has been noted earlier there are two important cropping seasons in Bihar viz. kharif season and rabi reason. Usually farmers grow winter rice, autumn rice, maize and ragi during the kharif season and wheat, barely and gram during the rabi season. All these crops except winter rice, the substitution possibilities of which are very much limited due to agro-climatic conditions required for growing winter rice, are included in the present study of crop-substitution for the districts of Santhal Pargana, Hazaribagh and Palamau but in other districts some crops which are grown in very small areas could not be included. The list of substitutable crops included in the present study in two different seasons in the districts of Chotanagpur region is given in Table 6.1.

Among the substitutable crops grown during the kharif season the greater proportion of area goes to maize as compared to autumn rice and ragi in the districts of Santhal Pargana, Hazaribagh, Palamau and Dhanbad in both the periods. But in the districts of Ranchi and Singhbhum a greater proportion of area is allotted to autumn rice as compared to maize and ragi in both the periods (see Table 6.2 for detail). Table 6.2 indicates that the average proportion of area under maize relative to that of autumn rice and ragi is greater in the HYV period than in the previous period in all the districts indicating that the farmers have preferred to grow maize as compared to autumn rice and ragi. This may be due to the introduction of HYV maize in the second period (i.e. HYV period).

Among the substitutable crops grown during the rabi season, we find that a greater proportion of area is given to gram as compared to wheat and barley in the pre-HYV period in all the districts but in the HYV period greater importance is given to the wheat crop as compared to barley and gram in the districts of Santhal Pargana, Hazaribagh and Ranchi (where the use of HYV wheat is relatively higher) whereas gram maintains its greater share of area in the districts of Palamau and Singhbhum as indicated by the average proportion of area under these crops in Table 6.2. Table 6.2 also indicates that the average proportion of area under wheat as compared to that of barley and gram has considerably increased in the

TABLE 6.2: **Showing ratio of average proportion of area under different crop combinations during kharif season and rabi season in the two sub-periods**

| District | Period | Ten years' average proportion* of area under different crop combinations during: KHARIF SEASON | | | RABI SEASON | | |
|---|---|---|---|---|---|---|---|
| | | Maize:Autumn -rice | Maize:Ragi | Autumn rice Ragi | Wheat:Gram | Wheat:Barley | Gram:Barley |
| Santhal-Pargana | Pre-HYV | 81:19 | 94:06 | 77:23 | 23:77 | 41:59 | 70:30 |
| | HYV | 86:14 | 95:05 | 75:25 | 66:34 | 75:25 | 62:38 |
| Hazari-bagh | Pre-HYV | 83:17 | 63:37 | 25:75 | 50:50 | 63:37 | 64:36 |
| | HYV | 83:17 | 67:33 | 29:71 | 74:26 | 79:21 | 57:43 |
| Palamau | Pre-HYV | 70:30 | 90:10 | 79:21 | 24:76 | 39:61 | 68:32 |
| | HYV | 77:23 | 92:08 | 77:23 | 44:56 | 52:48 | 58:42 |
| Ranchi | Pre-HYV | 08:92 | 16:84 | 69:31 | 29:71 | - | - |
| | HYV | 11:89 | 22:78 | 71:29 | 60:40 | - | - |
| Singhbhum | Pre-HYV | 33:67 | - | - | 19:81 | - | - |
| | HYV | 36:64 | - | - | 48:52 | - | - |
| Dhanbad | Pre-HYV | - | 68:32 | - | - | - | - |
| | HYV | - | 71:29 | - | - | - | - |

* Average proportions are computed from actual values of decadal ending 1965-66 for the pre-HYV period and of decadal ending 1976-77 for the HYV period.

HYV period in all the districts indicating that the farmers have substituted wheat for barley and gram in the HYV period. The highest increase in the average proportion of area under wheat relative to barley and gram is observed in Santhal Pargana where the percentage of area under HYV wheat is highest among the districts of Chotanagpur region.

## 6.3 FACTORS INFLUENCING FARMERS' DECISION REGARDING CROP-SUBSTITUTION IN CHOTANAGPUR REGION

The following are some of the factors which might affect farmers' decisions while allocating their land to different substitutable crops:

(1) Relative price of the crop might be an important factor influencing farmers' decisions regarding substitution of crops to the extent that they are price-responsive or market oriented. The relative price may be the only consideration for the crops grown for the market. But in the case of cereals which the traditional farmers mostly grow for their own consumption, the relative yield-rate by itself may also be an important consideration. It should be noted that if some crops are grown solely for their own consumption then their prices do not affect the farmers' decision in respect of allocation of land to them. Risk, quality and tradition might be working at the back of growing these crops solely for their own consumption. This usually applies to "small" farmers. As far as the larger farmers are concerned if they grow some crops for their own consumption it might be mostly for quality-attachment and tradition.

(2) Differences in the water requirements of the crops might also influence the farmers' decision for substituting one crop for another. For instance autumn rice requires relatively more rainfall water than maize or ragi, all the three being the kharif season crops. Under these agro-climatic conditions required for growing the crops the rational farmers are expected to grow autumn rice when sufficient rainfall occurs, and maize or ragi when rainfall becomes scanty. It is very probable that the farmers may have decided to plant autumn rice in a particular year but in the absence of sufficient rainfall during the sowing season they may not be able to do this and they may be compelled to change their decision and opt for maize or ragi

instead of autumn rice as desired by them. Again if irrigation facilities are given to the farmers, they are expected to substitute wheat for barley and gram during the rabi season. Substitution between the HYV and traditional variety wheat may also be there, subject to the availability of other inputs required for growing HYV wheat.

(3) The crop-rotation systems practised by the farmers as noted earlier in Chapter 2 may be a factor influencing substitution of one crop for another in alternative years. The crop-rotation system may be taken as a form of systematic crop-substitution which may be quite independent of price changes or at least short-run changes in relative price or returns in small magnitude may not be sufficient for inducing the farmers to change their tradition-dictated and time-tested crop-rotation practice. If this is true then one should not expect positive price response for the traditional farmers so far as their crop-substitution behaviour is concerned if these changes are not relatively large and persisting ones.

(4) The relative risk involved in growing the crops might be an important factor in the farmers' decision. A risk averter farmer might prefer to grow a crop which involves less risk, with regard to its price or return, as compared to other crops.

(5) If other things remain the same, the traditional farmers are expected to adjust the area in favour of those crops for which they are more trained and experienced. A farmer is here said to be more trained in growing crop A in relation to crop B (provided A and B are substitutable crops) if he has been allocating over the years a considerable proportion of area in growing crop A as compared to crop B.

In the next section we specify and later estimate the regression relationships between area substitution and other variables incorporating the points we have noted above. We are again using the working rule hypothesis discussed earlier. An explicit formal quantitative format leading to such a relationship is not difficult to build up [86] but we have not followed that line as our emphasis is here on the informal process actually followed rather than their theoretical formal counterpart.[1]

## 6.4 ALTERNATIVE SPECIFICATIONS OF THE REGRESSION EQUATIONS

It is quite possible that the traditional farmers, while deciding on any change in the area under the crop relative to its substitute (s), might consider the relative proportion of area given to the crop in the previous year and try to adjust the area in the current year in response to the changes in relative price/returns and some other factors which might influence their decision. Therefore, in our regression equations we have included relative areas of the pairs of crops in the immediately previous year, relative prices lagged one year, relative productivity lagged one year, relative gross return lagged one year and pre-sowing rainfall and irrigated area in the current year as independent variables. The ratio of the area under substitutable crops in the current year is taken as the dependent variable. The two regressions that have been chosen for estimation in all the districts and for all pairs of crops after a preliminary trying out of a number of relationships on the data of one district are

$$C.1 \ \mathrm{Log}\left(\frac{A_{rt}}{A_{st}} \times 100\right) = a + b_1 \mathrm{Log}\left(\frac{A_{rt-1}}{A_{st-1}} \times 100\right) + b_2 \mathrm{Log}\left(\frac{P_{rt-1}}{P_{st-2}} \times 100\right) + b_3 \mathrm{Log}\left(\frac{Y_{rt-1}}{Y_{st-1}} \times 100\right) + u_t$$

$$C.2 \ \mathrm{Log}\left(\frac{A_{rt}}{A_{st}} \times 100\right) = a + b_1 \mathrm{Log}\left(\frac{A_{rt-1}}{A_{st-1}} \times 100\right) + b_2 \mathrm{Log}\left(\frac{G_{rt-1}}{G_{st-2}} \times 100\right) + u_t$$

where $A_{rt}$, $P_{rt}$, $Y_{rt}$ and $G_{rt}$ are area in hectares, farm hervest price, yield-rate per hectare, gross return per hectare respectively of the rth crop in year t, (s standing for substitute crop) and $u_t$ is the random error term.

1. See foot note in section 1.6.

**TABLE 6.3: Showing cropping pattern response equations for pairs of crops grown during kharif season in the two sub-periods and in the combined period**

| Pairs of crops | Period | Eqn. No. | Constant | Coefficients of independent variable | | | | $R^2$ | $^F(V_1, V_2)$ |
|---|---|---|---|---|---|---|---|---|---|
| | | | | $\text{Log} \frac{A_{rt-1}}{A_{st-1}} \times 100$ | $\text{Log} \frac{P_{rt-1}}{P_{st-1}} \times 100$ | $\text{Log} \frac{Y_{rt-1}}{Y_{st-1}} \times 100$ | $\text{Log} \frac{G_{rt-1}}{G_{st-1}} \times 100$ | | |
| 1 | 2 | 3 | 4 | 5 | 6 | 7 | 8 | 9 | 10 |
| | | | | SANTHAL PARGANA DISTRICT | | | | | |
| <u>Maize</u> | Pre-HYV | C.1 | 2.578 | 0.458* | -0.031 | 0.166** | - | 0.746 | 5.879 ** |
| Autumn | | | (0.198) | (0.181) | (0.056) | | | (3,6) | |
| rice | | C.2 | 2.383 | 0.486** | - | - | 0.149** | 0.694 | 7.928 ** |
| | | | (0.200) | | | (0.055) | | (2,7) | |
| | HYV | C.1 | 10.688 | -0.804** | 0.061 | 0.120 | - | 0.638 | 3.528 * |
| | | | (0.248) | (0.370) | (0.093) | | | (3,6) | |
| | | C.2 | 10.956 | -0.807** | - | - | 0.127 | 0.637 | 6.132 ** |
| | | | | (0.232) | | | (0.082) | | (2,7) |
| | Comb- | C.1 | 2.062 | 0.272 | 0.528 | 0.016 | - | 0.332 | 2.645 * |
| | ined | | | (0.228) | (0.337) | (0.099) | | | (3,16) |
| | | C.2 | 3.580 | 0.452* | - | - | 0.030 | 0.207 | 2.219 |
| | | - | (0.215) | | | (0.101) | | (2,17) | |
| <u>Maize</u> | | | 7.205 | -0.553 | 1.122 | -0.239 | - | 0.126 | 0.288 |
| Ragi | | | (9.213) | (1.431) | (0.323) | | | (3,6) | |
| | | | 7.713 | -0.013 | - | - | -0.080 | 0.015 | 0.055 |
| | | | (0.021) | | | (0.256) | | (2,7) | |
| | | | 7.043 | 0.497 | -0.515 | -0.129 | - | 0.440 | 1.566 |
| | | | | (0.327) | (0.420) | (0.091) | | (3,63) | |

*Contd.* Table 6.3

| 1 | 2 | 3 | 4 | 5 | 6 | 7 | 8 | 9 | 10 | |
|---|---|---|---|---|---|---|---|---|---|---|
| | | | 5.650 | 0.372 | - | - | -0.158 | 0.371 | 2.063 | |
| | | | | (0.287) | | | (0.083) | | (2,7) | |
| | | | 2.175 | 0.401 | 0.733 | -0.245 | - | 0.247 | 1.751 | |
| | | | | (0.093) | (0.730) | (0.158) | | | (3,16) | |
| | | | 1.304 | 0.934 | - | - | -0.154 | 0.171 | 1.745 | |
| | | | | (0.551) | | | (0.142) | | (2,17) | |
| <u>Autumn rice</u> | | | | 7.727 | -0.064 | -0.345 | 0.032 | - | 0.213 | 0.540 |
| Ragi | | | | (0.327) | (0.302) | (0.089) | | | (3,6) | |
| | | | 5.246 | 0.091 | - | - | 0.008 | 0.017 | 0.062 | |
| | | | | (0.311) | | | (0.089) | | (2,7) | |
| | | | 6.648 | -0.200 | -0.165 | 0.195 | - | 0.056 | 0.119 | |
| | | | | (0.609) | (0.490) | (0.335) | | | (3,6) | |
| | | | 5.352 | 0.009 | - | - | 0.059 | 0.019 | 0.067 | |
| | | | | (0.407) | | | (0.174) | | (2,7) | |
| | | | 5.242 | 0.108 | -0.045 | 0.025 | - | 0.028 | 0.154 | |
| | | | | (0.262) | (0.237) | (0.101) | | | (2,16) | |
| | | | 4.931 | 0.133 | - | - | 0.013 | 0.024 | 0.208 | |
| | | | | (0.236) | | | (0.087) | | (2,17) | |
| | | | | HAZARIBAGH DISTRICT | | | | | | |
| <u>Maize</u> | | | 11.583 | -0.585 | -0.245* | -0.127 | - | 0.719 | 5.139 ** | |
| Autumn rice | | | | | (0.317) | (0.105) | (0.104) | | | (3,6) |
| | | | 9.886 | -0.442 | - | - | -0.186*** | 0.701 | 8.201 ** | |
| | | | | (0.214) | | | (0.048) | | (2,7) | |
| | | | 9.510 | -0.468 | -0.015 | -0.058 | - | 0.477 | 1.824 | |
| | | | | (0.352) | (0.094) | (0.042) | | | (2,6) | |
| | | | 8.890 | -0.380 | - | - | -0.056 | 0.450 | 2.865 | |
| | | | | (0.292) | | | (0.041) | | (2,7) | |

*Contd.* Table 6.3

| 1 | 2 | 3 | 4 | 5 | 6 | 7 | 8 | 9 | 10 |
|---|---|---|---|---|---|---|---|---|---|
| | | | 9.520 | -0.335* | -0.131** | -0.119** | - | 0.504 | 5.433 *** |
| | | | | (0.186) | (0.054) | (0.047) | | | (3,16) |
| | | | 8.875 | -0.324* | - | - | -0.124*** | 0.503 | 8.576 *** |
| | | | | (0.171) | | | (0.033) | | (2,17) |
| <u>Maize</u> | | | 4.425 | 0.251 | -0.159 | 0.034 | - | 0.429 | 1.499 |
| Ragi | | | | (0.449) | (0.206) | (0.034) | - | | (3,6) |
| | | | 2.197 | 0.545 | - | - | 0.025 | 0.350 | 1.885 |
| | | | | (0.307) | | | (0.032) | | (2,7) |
| | | | 3.962 | 0.143 | 0.122 | -0.004 | - | 0.201 | 0.501 |
| | | | | (0.433) | (0.197) | (0.083) | | | (3,6) |
| | | | 3.864 | 0.247 | - | - | 0.067 | 0.165 | 0.691 |
| | | | | (0.363) | | | (0.066) | | (2,7) |
| | | | 1.756 | 0.718*** | -0.089 | 0.028 | - | 0.624 | 8.881 *** |
| | | | | (0.163) | (0.102) | (0.039) | | | (3,16) |
| | | | 1.259 | 0.750*** | - | - | 0.010 | 0.601 | 12.75 *** |
| | | | | (0.160) | | | (0.035) | | (2,17) |
| <u>Ragi</u> | | | 9.293 | -0.688*** | -0.180* | 0.220** | - | 0.759 | 6.267 ** |
| Autumn rice | | | | | (0.201) | (0.083) | (0.086) | | (3,6) |
| | | | 7.843 | -0.384 | - | - | 0.009 | 0.152 | 0.628 |
| | | | | (0.358) | | | (0.115) | | (2,7) |
| | | | 5.229 | -0.026 | 0.118 | -0.026 | - | 0.335 | 1.008 |
| | | | | (0.311) | (0.071) | (0.051) | | | (3,6) |
| | | | 5.542 | -0.028 | - | - | 0.027 | 0.077 | 0.291 |
| | | | | (0.340) | | | (0.042) | | (2,7) |
| | | | 6.294 | -0.135 | -0.122 | 0.127* | - | 0.265 | 1.928 |
| | | | | (0.247) | (0.075) | (0.071) | | | (3,16) |

*Contd.* Table 6.3

| 1 | 2 | 3 | 4 | 5 | 6 | 7 | 8 | 9 | 10 |
|---|---|---|---|---|---|---|---|---|---|
| | | | 4.757 | -0.144<br>(0.240) | - | - | 0.009<br>(0.055) | 0.023 | 0.205<br>(2,17) |
| | | | | PALAMAU DISTRICT | | | | | |
| Maize<br>Autmn rice | | | 2.892 | 0.436<br>(0.509) | 0.039<br>(0.231) | 0.002<br>(0.158) | - | 0.170 | 0.408<br>(3,6) |
| | | | 3.336 | 0.385<br>(0.327) | - | - | 0.003<br>(0.125) | 0.166 | 0.695<br>(2,7) |
| | | | 7.479 | -0.136<br>(0.317) | -0.320<br>(0.327) | 0.112<br>(0.103) | - | 0.211 | 0.535<br>(3,6) |
| | | | 6.865 | -0.274<br>(0.303) | - | - | 0.110<br>(0.082) | 0.210 | 0.932<br>(2,7) |
| | | | 1.343 | 0.708***<br>(0.199) | 0.127<br>(0.150) | -0.047<br>(0.097) | - | 0.471 | 4.750 **<br>(3,16) |
| | | | 1.797 | 0.674***<br>(0.197) | - | - | 0.010<br>(0.088) | 0.436 | 6.578 ***<br>(2,17) |
| Maize<br>Ragi | | | 6.049 | 0.132<br>(0.290) | -0.009<br>(0.108) | -0.023<br>(0.073) | - | 0.036 | 0.076<br>(3,6) |
| | | | 6.034 | 0.126<br>(0.267) | - | - | -0.021<br>(0.066) | 0.033 | 0.120<br>(2,7) |
| | | | 7.337 | 0.209<br>(0.320) | -0.213<br>(0.293) | -0.144<br>(0.148) | - | 0.268 | 0.733<br>(3,6) |
| | | | 6.123 | 0.240<br>(0.282) | - | - | -0.140<br>(0.138) | 0.258 | 1.218<br>(2,7) |
| | | | 3.292 | 0.660***<br>(0.159) | -0.093<br>(0.151) | -0.093<br>(0.086) | - | 0.564 | 6.882 ***<br>(3,16) |
| | | | 2.867 | 0.659***<br>(0.149) | - | - | -0.093<br>(0.083) | 0.564 | 10.99 ***<br>(2,17) |

*Contd.* Table 6.3

| 1 | 2 | 3 | 4 | 5 | 6 | 7 | 8 | 9 | 10 | |
|---|---|---|---|---|---|---|---|---|---|---|
| Autumn rice | | | | 4.168 | 0.485 | -0.051 | -0.184 | - | 0.371 | 1.181 |
| Ragi | | | | (0.379) | (0.209) | (0.220) | | | (3,6) | |
| | | | 4.313 | 0.375 | - | - | -0.117 | 0.365 | 2.011 | |
| | | | | (0.242) | | | (0.131) | | (2,7) | |
| | | | 2.841 | 0.422 | -0.195 | 0.317 | - | 0.477 | 1.828 | |
| | | | | (0.366) | (0.196) | (0.184) | | | (3,6) | |
| | | | 2.673 | 0.483 | - | - | 0.073 | 0.233 | 1.067 | |
| | | | | (0.408) | | | (0.127) | | (2,7) | |
| | | | 2.902 | 0.439* | 0.001 | 0.088 | - | 0.269 | 1.966 | |
| | | | | (0.231) | (0.111) | (0.138) | | | (3,16) | |
| | | | 2.900 | 0.485** | - | - | 0.029 | 0.256 | 2.930 * | |
| | | | | (0.204) | | | (0.080) | | (2,17) | |
| | | | | | RANCHI DISTRICT | | | | | |
| Maize | | | 2.465 | -0.122 | -0.017 | 0.001 | - | 0.026 | 0.054 | |
| Autumn rice | | | | | (0.643) | (0.084) | (0.119) | | | (3,6) |
| | | | 2.362 | -0.081 | - | - | -0.011 | 0.024 | 0.087 | |
| | | | | (0.504) | | | (0.066) | | (2,7) | |
| | | | 4.848 | -0.278 | -0.263 | -0.090 | - | 0.256 | 0.690 | |
| | | | | (0.300) | (0.336) | (0.171) | | | (3,6) | |
| | | | 3.714 | -0.329 | - | - | -0.084 | 0.209 | 0.924 | |
| | | | | (0.275) | | | (0.163) | | (2,7) | |
| | | | 1.443 | 0.522** | 0.017 | -0.078 | - | 0.311 | 2.407 | |
| | | | | (0.231) | (0.170) | (0.124) | | | (3,16) | |
| | | | 1.267 | 0.563** | - | - | -0.048 | 0.300 | 3.646 ** | |
| | | | | (0.210) | | | (0.106) | | (2,17) | |

*Contd.* Table 6.3

| 1 | 2 | 3 | 4 | 5 | 6 | 7 | 8 | 9 | 10 | |
|---|---|---|---|---|---|---|---|---|---|---|
| Maize | | | 3.173 | -0.110 | 0.116 | -0.089 | - | 0.118 | 0.267 | |
| Ragi | | | | (0.396) | (0.251) | (0.135) | | | (3,6) | |
| | | | 3.243 | -0.005 | - | - | -0.054 | 0.028 | 0.100 | |
| | | | | (0.362) | | | (0.124) | | (2,7) | |
| | | | 6.606 | -0.322 | -0.254 | -0.192 | - | 0.432 | 1.521 | |
| | | | | (0.211) | (0.308) | (0.175) | | | (3,6) | |
| | | | 5.422 | -0.335 | - | - | -0.183 | 0.425 | 2.584 | |
| | | | | (0.192) | | | (0.160) | | (2,7) | |
| | | | 3.202 | 0.609*** | -0.264 | -0.139 | - | 0.401 | 3.575 ** | |
| | | | | (0.195) | (0.255) | (0.152) | | | (3,16) | |
| | | | 2.070 | 0.609*** | - | - | -0.158 | 0.392 | 5.464 ** | |
| | | | | (0.191) | | | (0.143) | | (2,17) | |
| Autumn rice | | | | 3.821 | 0.292 | 0.019 | -0.018 | - | 0.126 | 0.288 |
| Ragi | | | | (0.370) | (0.131) | (0.233) | | | (3,6) | |
| | | | 3.674 | 0.311 | - | - | 0.011 | 0.123 | 0.490 | |
| | | | | (0.321) | | | (0.112) | | (2,7) | |
| | | | 1.724 | 1.052** | -0.210 | -0.228 | - | 0.542 | 2.366 | |
| | | | | (0.401) | (0.179) | (0.141) | | | (2,6) | |
| | | | 0.796 | 1.048** | - | - | -.0.231 | 0.539 | 4.096 * | |
| | | | | (0.371) | | | (0.130) | | (2,7) | |
| | | | 3.389 | 0.420* | -0.033 | -0.033 | - | 0.198 | 1.314 | |
| | | | | (0.214) | (0.077) | (0.101) | | | (3,16) | |
| | | | 3.280 | 0.422* | - | - | -0.028 | 0.196 | 2.074 | |
| | | | | (0.208) | | | (0.071) | | (2,17) | |
| SINGHBHUM DISTRICT | | | | | | | | | | |
| Maize | | | 1.953 | 0.639 | -0.181 | 0.064 | - | 0.349 | 1.076 | |
| Autumn rice | | | | | (0.383) | (0.316) | (0.316) | | | (3,6) |

*Contd.* Table 6.3

| 1 | 2 | 3 | 4 | 5 | 6 | 7 | 8 | 9 | 10 |
|---|---|---|---|---|---|---|---|---|---|
| | | | 1.796 | 0.622 | - | - | -0.052 | 0.324 | 1.674 |
| | | | | (0.359) | | | (0.166) | | (2,7) |
| | | | 0.314 | 0.706*** | 0.169 | 0.011 | - | 0.815 | 8.885 ** |
| | | | | (0.161) | (0.101) | (0.040) | | | (3,6) |
| | | | 1.275 | 0.620*** | - | - | 0.044 | 0.764 | 11.300*** |
| | | | | (0.155) | | | (0.033) | | (2,7) |
| | | | 1.892 | 0.602** | -0.116 | 0.045 | - | 0.408 | 3.681 ** |
| | | | | (0.207) | (0.171) | (0.118) | | | (3,16) |
| | | | 1.721 | 0.580*** | - | - | -0.013 | 0.391 | 5.447 ** |
| | | | | (0.202) | | | (0.083) | | (2,17) |
| | | | | DHANBAD DISTRICT | | | | | |
| Maize / Ragi | | | 4.250 | -0.006 | 0.340 | -0.101 | - | 0.113 | 0.255 |
| | | | | (0.485) | (0.653) | (0.248) | | | (3,6) |
| | | | 4.866 | 0.184 | - | - | -0.089 | 0.038 | 0.137 |
| | | | | (0.391) | | | (0.240) | | (2,7) |
| | | | 2.217 | 0.565* | -0.018 | 0.055 | - | 0.759 | 6.265 ** |
| | | | | (0.236) | (0.072) | (0.030) | | | (3,6) |
| | | | 2.017 | 0.586** | - | - | 0.052 | 0.711 | 8.583 ** |
| | | | | (0.238) | | | (0.031) | | (2,17) |
| | | | 3.749 | 0.322 | -0.030 | 0.019 | - | 0.106 | 0.631 |
| | | | | (0.256) | (0.225) | (0.089) | | | (3,16) |
| | | | 3.668 | 0.310 | - | - | 0.018 | 0.103 | 0.974 |
| | | | | (0.243) | | | (0.087) | | (2,17) |

The above equations are used for studying the general behaviour of crop-substitution in the Chotanagpur region considering all possible pairs of crops grown during kharif season and rabi season separately. Keeping the characteristics of the crops grown in the two cropping seasons in mind we have regressed log $(A_{rt}/A_{st} \times 100)$ on Log $R_{rt}$, $R_{rt}$ being the pre-sowing rainfall relevant to the rth crop in year t for examining the importance of pre-sowing rainfall in the substitution of area between kharif crops and on Log $I_t$, $I_t$ being the total irrigated area during rabi season in year t, for the purpose of studying the importance of irrigation facilities in acreage substitution between rabi crops. We refer to these regressions as $C_3$ and $C_4$ in the discussion and tables that follow.

It should be noted that all the factors influencing farmers' decision which were discussed in the earlier sections, could not be included in our regression equations due to the non-availability of relevant time-series data. Allowing sufficient degrees of freedom with the time series data available for analysis in each of the periods also acted as a constraint in introducing many regressors in the regression equations. However, we have kept these factors, specially crop-rotation system and relative risk[2] and their impacts in mind while interpreting the coefficients of the variables included in the regressions.

## 6.5 REGRESSION RESULTS

Keeping in mind the factors briefly discussed above, we examine the crop-substitution behaviour of the farmers as revealed by our data. The log-linear regression results obtained (using equation $C_1$ and $C_2$) for different pairs of crops for each of the districts of Chotanagpur region in the two sub-periods and also for the combined period and for the two important cropping seasons separately are given below in Tables 6.3 and 6.4 for kharif season and rabi season respectively. We have also computed the F-ratios associated with the test of overall homogeneity of regressions in the two sub-periods and the results are presented in Table 6.5.[3] Since our regres-

---

2. Risk is here measured by the coefficient of variation of gross return per hectare. Detailed results indicating the magnitude of risk involved in growing different substitutable crops in the two sub-periods are given in Table 6.8

sions are linear in logarithms, the coefficients of the independent variable provide a direct and readymade estimate of the constant elasticity of substitution between the crops which may be interpreted as in single product production function analysis [86 p.140] but here applied to substitution between outputs as represented by acreages (see for detail Appendix VI at the end of this chapter).

We take up the discussion of results season by season. We first take up the case of kharif season and then the case of rabi season.

### 6.5.1 Kharif Season

Regression results are presented in Table 6.3. We find that the farmers of Santhal Pargana district have substituted maize for autumn rice in both the periods but the factors influencing their decision appear to be different from period to period. There is no price responsiveness but responsiveness to yield and tradition variables is there in the pre-HYV period as indicated by the positive and significant coefficients of the yield and lagged dependent variables. In the HYV period the yield coefficient becomes insignificant (though positive) but this may be due to greater uncertainty and greater fluctuations prevailing in that period. The lagged dependent variable comes out significant with negative sign which might indicate the adoption of traditional crop-rotation system which is followed perhaps for uncertainty coverage and better land-use. No such substitution between maize and ragi and autumn rice and ragi is indicated but this may be because of the fact that ragi is a minor crop and mainly grown by the poor farmers for their own consumption. The Chow test indicates that the substitution-behaviour between maize and autumn rice has significantly changed in the two sub-periods ( see Table 6.5).

The Hazaribagh farmers seem to be indifferent towards economic considerations in planting maize and autumn rice. The coefficients of relative price and relative return variable are negative and significant in the pre-HYV period and also for the combined period. However, this may be due to the system of crop-rotation

---

3. It should be noted that t and F tests have only asymptotic validity due to the presence of lagged endogenous variable in the regressions.

practised by the farmers. This is also borne out by the negative and significant coefficients of the lagged dependent variable. The farmers have, however, adjusted the area in favour of maize as against ragi. This may be because of the increase in the risk of growing ragi as compared to maize. The coefficient of variation of gross return of ragi has considerably increased relatively to that of maize in the HYV period (see Table 6.8). The crop-substitution behaviour remains unchanged in the two sub-periods.

Palamau farmers are more traditional in planting maize, autumn rice and ragi. For, except the coefficient of lagged dependent variable in the combined period, no other coefficients are significant. $R^2$ values are very small in all the cases indicating that some other important factors are dominating in the farmers' decisions. The crop substitution behaviour remains almost the same in the two sub-periods. Ranchi farmers are also traditional in growing maize, autumn rice and ragi and going on planting these crops in alternative years. They seem to be indifferent towards the economic considerations in planting these crops as indicated by the negative coefficients of relative price and relative return variables. But this may be due to the system of crop-rotation practised by the farmers as noted earlier and it is also supported by the negative coefficients of the lagged dependent variables in the two sub-periods. Surprisingly the coefficients of the lagged dependent variable comes out positive and significant in the combined period regression. This may be due to sudden increase in the area under maize relative to autumn rice and ragi in the HYV period due to a widespread use of HYV maize in the district (The percentage of area under HYV maize is highest in Ranchi among the Chotanagpur districts).

The farmers of Singhbhum and Dhanbad districts have traditionally substituted maize for autumn rice and ragi respectively specially in the HYV period. Except the coefficient of lagged dependent variable in the HYV period no other coefficients are significant. However, no significant change in the crop substitution behaviour between the two sub-periods is indicated in these districts.

The $R^2$'s in all the equations are not very high, indicating, as has been noted in the earlier discussions, that there are other factors influencing the crop-substitution decisions of the farmers. As has been noted earlier, presowing rainfall is one such factor in the kharif season. The importance of pre-sowing rainfall is separately analysed by regressing the logarithm of the ratio of area under the crops in percentage on the logarithm of the pre-sowing rainfall relevant to the crops. Regression equations have been estimated for the pairs of maize & autumn rice and ragi & autumn rice (because maize and ragi requires less water as compared to autumn rice as noted earlier) for all relevant districts but only those regressions which give significant coefficients for pre-sowing rainfall have been presented. These regressions are given in Table 6.6. We find that the pre-sowing rainfall coefficients give negative sign as expected and are also significant in the HYV period in the districts of Santhal Pargana, Palamau and Ranchi indicating the importance of pre-sowing rainfall in the acreage substitution between maize and autumn rice. The difference in the water requirements of these crops are also indicated by the coefficients of the pre-sowing rainfall variable in our acreage response-equations examined earlier in Chapter 5, where we find positive coefficients for autumn rice but negative coefficients for maize and ragi.

It should be noted that the coefficients of pre-sowing rainfall variable in Table 6.6 are significant only in the HYV period and not in the pre-HYV period. This might indicate that the farmers have utilised the scanty rainfall available during the sowing season in a better way in the HYV period, when the average value of the pre-sowing rainfall relevant to these crops (i.e. Bhadai crops) has invariably declined and the coefficient of variation (emplying greater weather uncertainty) has gone up in the HYV period in the districts of Chotanagpur region as compared to the pre-HYV period, as noted earlier in Chapter 3. It should also be noted that the $R^2$ value in the simple regression with only the pre-sowing rainfall variable, exceeds the $R^2$ value obtained in the multiple regressions (i.e. equation C.1 and C.2) during the HYV period in the districts of Palamau and Ranchi indicating the dominance of the weather factor in acreage substitution decisions of the farmers.

Thus we find that the farmers have substituted maize (for which the HYV seeds are available) for autumn rice and ragi in the HYV period in all the districts but they are not price responsive. Weather conditions and tradition dominate over the price considerations in

TABLE 6.4: **Showing cropping pattern response equations for pairs of crops grown during rabi season in the two sub-periods and in the combined period**

| Pairs of crops | Period | Eqn. No. | Constant | Coefficients of independent variable | | | | $R^2$ | $^F(V_1, V_2)$ |
|---|---|---|---|---|---|---|---|---|---|
| | | | | $\text{Log}\frac{A_{rt-1}}{A_{st-1}}\times 100$ | $\text{Log}\frac{P_{rt-1}}{P_{st-1}}\times 100$ | $\text{Log}\frac{Y_{rt-1}}{Y_{st-1}}\times 100$ | $\text{Log}\frac{G_{rt-1}}{G_{st-1}}\times 100$ | | |
| 1 | 2 | 3 | 4 | 5 | 6 | 7 | 8 | 9 | 10 |
| | | | | SANTHAL PARGANA DISTRICT | | | | | |
| Wheat | Pre-HYV | C.1 | -5.736 | 0.478 (0.386) | 1.037 (0.813) | 0.539* (0.232) | - | 0.486 | 1.889 (3,6) |
| Gram | | C.2 | -1.719 | 0.635 (0.368) | - | - | 0.600** (0.214) | 0.536 | 4.031 (2,7) |
| | HYV | C.1 | 4.692 | 0.390* (0.193) | -0.456 (0.763) | 0.124 (0.257) | - | 0.696 | 4.583 * (3,6) |
| | | C.2 | 2.223 | 0.485** (0.145) | - | - | 0.105 (0.248) | 0.666 | 6.977 ** (2,7) |
| | Combined | C.1 | -3.201 | 0.862*** (0.136) | 0.343 (0.837) | 0.454* (0.224) | - | 0.878 | 38.28*** (3,16) |
| | | C.2 | -1.756 | 0.880*** (0.097) | - | - | 0.463* (0.213) | 0.879 | 61.81 *** (2,17) |
| Wheat | | | 1.217 | 0.318 (0.348) | 0.093 (0.453) | 0.261 (0.312) | - | 0.154 | 0.363 (3,6) |
| Barley | | | 1.752 | 0.315 (0.327) | - | - | 0.231 (0.272) | 0.141 | 0.575 (2,7) |
| | | | -1.374 | 0.874*** (0.144) | 0.141 (1.017) | 0.306 (0.474) | - | 0.889 | 15.99*** (3,6) |

*Contd. Table 6.4*

| 1 | 2 | 3 | 4 | 5 | 6 | 7 | 8 | 9 | 10 |
|---|---|---|---|---|---|---|---|---|---|
| | | | 0.925 | 0.891 | - | - | 0.318 | 0.887 | 27.72*** |
| | | | | (0.120) | - | - | (0.429) | | (2,7) |
| | | | -3.495 | 0.896*** | 0.242 | 0.587*** | - | 0.916 | 58.12*** |
| | | | | (0.096) | (0.406) | (0.188) | | | (3,16) |
| | | | -2.604 | 0.929*** | - | - | 0.577*** | 0.914 | 90.76*** |
| | | | | (0.087) | | | (0.184) | | (2,17) |
| Gram | | | 5.244 | -0.351 | 0.139 | 0.330 | - | 0.462 | 1.720 |
| | | | | (0.087) | (0.246) | (0.226) | | | (3,6) |
| Barley | | | 6.210 | -0.344 | - | - | 0.245 | 0.422 | 2.564 |
| | | | | (0.275) | | | (0.179) | | (2,7) |
| | | | -2.689 | 0.386 | 1.337 | -0.168 | - | 0.489 | 1.913 |
| | | | | (0.374) | (1.156) | (0.315) | | | (3,6) |
| | | | 1.663 | 0.638 | - | - | 0.035 | 0.376 | 2.106 |
| | | | | (0.311) | | | (0.263) | | (2,7) |
| | | | 0.401 | 0.611*** | 0.196 | 0.152 | - | 0.388 | 3.383 ** |
| | | | | (0.193) | (0.350) | (0.227) | | | (3,16) |
| | | | 1.270 | 0.609*** | - | - | 0.163 | 0.387 | 5.371 |
| | | | | (0.186) | | | (0.190) | | (2,17) |
| HAZARIBAGH DISTRICT | | | | | | | | | |
| Wheat | | | 6.400 | -0.072 | -0.468 | -0.154 | - | 0.132 | 0.303 |
| | | | | (0.435) | (0.720) | (0.284) | | | (3,6) |
| Gram | | | 5.003 | -0.139 | - | - | -0.047 | 0.047 | 0.173 |
| | | | | (0.414) | | | (0.241) | | (2,7) |
| | | | 7.780 | 0.015 | -0.751 | 0.206 | - | 0.432 | 1.523 |
| | | | | (0.414) | (0.662) | (0.300) | | | (3,6) |

*Contd. Table 6.4*

| 1 | 2 | 3 | 4 | 5 | 6 | 7 | 8 | 9 | 10 |
|---|---|---|---|---|---|---|---|---|---|
| | | | 2.442 | 0.422 | - | - | 0.162 | 0.204 | 0.899 |
| | | | | (0.351) | | | (0.327) | | (2,7) |
| | | | 3.453 | 0.604*** | -0.642 | 0.310 | - | 0.746 | 15.72 *** |
| | | | | (0.201) | (0.518) | (0.199) | | | (3,16) |
| | | | -0.775 | 0.889*** | - | - | 0.276 | 0.687 | 18.73 *** |
| | | | - | (0.145) | | | (0.214) | | (2,17) |
| Wheat | | | 0.502 | 0.054 | 1.039 | -0.163 | - | 0.123 | 0.279 |
| | | | | (0.471) | (1.176) | (0.276) | | | (3,6) |
| Barley | | | 5.311 | -0.011 | - | - | -0.015 | 0.072 | 0.003 |
| | | | | (0.460) | | | (0.221) | | (2,7) |
| | | | -1.895 | 1.092*** | 0.114 | 0.185 | - | 0.821 | 9.130 *** |
| | | | | (0.255) | (0.425) | (0.171) | | | (2,6) |
| | | | -1.237 | 1.124*** | - | - | 0.122 | 0.801 | 14.07 *** |
| | | | | (0.233) | | | (0.142) | | (2,7) |
| | | | -1.239 | 0.884*** | 0.152 | 0.247* | - | 0.712 | 13.17 *** |
| | | | | (0.198) | (0.493) | (0.140) | | | (3,16) |
| | | | -0.513 | 0.904*** | - | - | 0.215* | 0.702 | 20.03 *** |
| | | | | (0.159) | | | (0.123) | | (2,17) |
| Gram | | | 2.511 | 0.500 | -0.007 | 0.019 | - | 0.350 | 1.078 |
| | | | | (0.579) | (0.296) | (0.189) | | | (3,6) |
| Barley | | | 2.544 | 0.493 | - | - | 0.012 | 0.349 | 1.881 |
| | | | | (0.529) | | | (0.152) | | (2,7) |
| | | | 3.109 | -0.030 | 0.365 | 0.027 | - | 0.334 | 1.003 |
| | | | | (0.340) | (0.216) | (0.168) | | | (3,6) |
| | | | 4.540 | -0.065 | - | - | 0.137 | 0.092 | 0.356 |
| | | | | (0.366) | | | (0.163) | | (2,7) |

Contd. Table 6.4

| 1 | 2 | 3 | 4 | 5 | 6 | 7 | 8 | 9 | 10 |
|---|---|---|---|---|---|---|---|---|---|
| | | | 1.362 | 0.603**** | 0.184 | -0.071 | - | 0.493 | 5.186 ** |
| | | | | (0.187) | (0.176) | (0.106) | | | (3,16) |
| | | | 2.075 | 0.589 | - | - | -0.005 | 0.441 | 6.717 *** |
| | | | | (0.190) | | | (0.095) | | (2,17) |
| | | | | PALAMAU DISTRICT | | | | | |
| Wheat | | | 9.120 | -0.084 | -0.921 | -0.208 | - | 0.694 | 4.549 |
| ——— | | | | (0.185) | (0.309) | (0.106) | | | (3,6) |
| Gram | | | 4.923 | -0.018 | - | - | -0.278 | 0.448 | 2.839 |
| | | | | (0.227) | | | (0.126) | | (2,7) |
| | | | 12.470 | -0.278 | -1.509 | 0.020 | - | 0.285 | 0.799 |
| | | | | (0.349) | (0.987) | (0.201) | | | (3,6) |
| | | | 4.453 | 0.011 | - | - | -0.016 | 0.001 | 0.004 |
| | | | | (0.322) | | | (0.216) | | (2,7) |
| | | | 4.894 | 0.664*** | -0.847 | 0.087 | - | 0.687 | 11.68 *** |
| | | | | (0.175) | (0.727) | (0.179) | | | (3,16) |
| | | | 6.586 | 0.798*** | - | - | 0.040 | 0.656 | 16.17 *** |
| | | | | (0.141) | | | (0.177) | | (2,17) |
| Wheat | | | 5.948 | 0.556* | -0.795 | -0.036 | - | 0.570 | 2.646 |
| ——— | | | | (0.283) | (0.435) | (0.126) | | | (3,6) |
| Barley | | | 1.855 | 0.601 | - | - | -0.033 | 0.329 | 1.717 |
| | | | | (0.326) | | | (0.146) | | (2,7) |
| | | | 1.147 | 0.925* | -0.269 | 0.117 | - | 0.709 | 4.877 ** |
| | | | | (0.426) | (0.465) | (0.126) | | | (3,6) |
| | | | -0.622 | 1.073** | - | - | 0.066 | 0.684 | 7.548 ** |
| | | | | (0.361) | | | (0.101) | | (2,7) |

*Contd. Table 6.4*

| 1 | 2 | 3 | 4 | 5 | 6 | 7 | 8 | 9 | 10 |
|---|---|---|---|---|---|---|---|---|---|
| | | | 2.259 | 0.902*** | -0.448 | 0.089 | - | 0.830 | 20.03 *** |
| | | | | (0.134) | (0.292) | (0.079) | | | (3,16) |
| | | | 0.052 | 0.935*** | - | - | 0.057 | 0.796 | 33.20*** |
| | | | | (0.141) | | | (0.081) | | (2,17) |
| Gram | | | 6.132 | 0.315 | -0.699 | 0.203 | | 0.526 | 2.214 |
| | | | | (0.236) | (0.336) | (0.107) | | | (3,6) |
| Barley | | | 3.605 | 0.294 | - | - | 0.036 | 0.129 | 0.517 |
| | | | | (0.294) | | | (0.095) | | (2,7) |
| | | | 4.243 | 0.287 | -0.071 | -0.097 | - | 0.098 | 0.381 |
| | | | | (0.456) | (1.995) | (0.381) | | | (3,6) |
| | | | 3.943 | 0.283 | - | - | -0.097 | 0.098 | 0.381 |
| | | | | (0.325) | | | (0.353) | | (2,7) |
| | | | 5.705 | 0.485* | -0.591 | -0.052 | - | 0.480 | 4.93 ** |
| | | | | (0.237) | (0.758) | (0.179) | | | (3,16) |
| | | | 2.625 | 0.584*** | - | - | -0.113 | 0.466 | 7.45 *** |
| | | | | (0.176) | | | (0.148) | | (2,17) |
| **RANCHI DISTRICT** | | | | | | | | | |
| Wheat | | | 1.527 | 0.678 | -0.552 | 0.461* | - | 0.689 | 4.423 * |
| | | | | (0.362) | (0.331) | (0.202) | | | (3,6) |
| Gram | | | -0.130 | 0.625 | - | - | 0.301 | 0.162 | 0.677 |
| | | | | (0.549) | | | (0.298) | | (2,7) |
| | | | -4.104 | 0.669** | 0.792 | 0.404 | - | 0.962 | 52.00 *** |
| | | | | (0.188) | (0.402) | (0.172) | | | (3,6) |
| | | | -0.215 | 0.655** | - | - | 0.363* | 0.955 | 73.96 *** |
| | | | | (0.192) | | | (0.172) | | (2,7) |

*Contd. Table 6.4*

| 1 | 2 | 3 | 4 | 5 | 6 | 7 | 8 | 9 | 10 |
|---|---|---|---|---|---|---|---|---|---|
| | | | 0.450 | 0.810*** | -0.208 | 0.267** | - | 0.939 | 81.67 *** |
| | | | | (0.102) | (0.299) | (0.101) | | | (3,16) |
| | | | -0.855 | 0.913*** | - | - | 0.251** | 0.929 | 110.4 *** |
| | | | | (0.084) | | | (0.105) | | (2,17) |
| | | | | SINGHBHUM DISTRICT | | | | | |
| Wheat | | | 0.210 | 0.538 | 0.115 | 0.139 | - | 0.352 | 1.083 |
| | | | | (0.313) | (0.356) | (0.210) | | | (3,6) |
| Gram | | | 0.768 | 0.534 | - | - | 0.135 | 0.350 | 1.889 |
| | | | | (0.284) | | | (0.189) | | (2,7) |
| | | | 2.954 | 0.959*** | -0.428 | -0.122 | - | 0.918 | 22.57 *** |
| | | | | (0.159) | (0.953) | (0.226) | | | (3,6) |
| | | 0.812 | 0.982*** | - | - | -0.102 | 0.918 | 38.63 *** | |
| | | | | (0.135) | | | (0.205) | | (2,7) |
| | | | 1.762 | 0.990*** | -0.318 | -0.024 | - | 0.929 | 69.48 *** |
| | | | | (0.095) | (0.316) | (0.142) | | | (3,16) |
| | | | 0.181 | 1.045*** | - | - | -0.043 | 0.924 | 104.1 *** |
| | | | | (0.076) | | | (0.140) | | (2,17) |

the farmers' decision. Negative relative price response observed for Hazaribagh and Ranchi districts is quite consistent with the crop-rotation system practised by the farmers of these districts. On the whole, non-price factors appear to be more important than the price factor in influencing the acreage substitution decisions of the Chotanagpur farmers.

### 6.5.2 Rabi Season

Regression results are presented in Table 6.4. We find that the farmers of Santhal Pargana district have substituted wheat for gram and barley, specially in the HYV period. The coefficients of the lagged dependent variable, rather than relative price coefficients, are significant. Farmers have, however, responded to relative yield-rate changes and changes is the relative returns. Relative yield-rate and relative return coefficients are significantly positive in the pre-HYV period for the pair wheat & gram and both for wheat & gram and wheat & barley in the combined period regression. No such substitution between barley and gram is indicated. This may be due to the inter cropping system. Farmers are reported to grow barley and gram simultaneously on the same plot of land during the rabi season as noted earlier in Chapter 2. The overall homogeneity of regressions in the two sub-periods for the pair, wheat & gram, is rejected at 1 percent level of significance (detailed results are given in Table 6.5).

TABLE 6.5:: **Showing computed values of F-ratios in testing for the overall homogeneity of cropping pattern response equations in the two sub-periods during the kharif season and rabi season (Eqn. No. C.1 and degrees of freedom is $V_1 = 4$ & $V_2 = 12$)**

| District | Computed F-values for different pairs of competing crops grown during | | | | | |
|---|---|---|---|---|---|---|
| | KHARIF SEASON | | | RABI SEASON | | |
| | Maize & Autumn rice | Maize & Ragi | Autumn rice & Ragi | Wheat & Gram | Wheat & Barley | Gram & Barley |
| Santhan Pargana | 13.182*** | 0.447 | 0.513 | 6.546*** | 0.958 | 2.804* |
| Hazaribagh | 1.675 | 1.317 | 8.022*** | 1.044 | 1.784 | 2.363 |
| Palamau | 2.541* | 1.992 | 1.092 | 6.105*** | 0.895 | 1.303 |
| Ranchi | 6.484*** | 6.632*** | 0.440 | 3.974* | - | - |
| Singhbhum | 0.111 | - | - | 1.188 | - | - |
| Dhanbad | - | 0.917 | - | - | - | - |

In the case of Hazaribagh district, we find that the farmers have traditionally substituted wheat for gram and barley. Except for the lagged dependent variable, no other variables have significant coefficients. $R^2$ values are, however, very small indicating that some other important factors are operating behind the decision of the farmers. Chow test indicates that the crop-substitution-behaviour remains almost the same in the two sub-periods. The acreage substitution-behaviour of the Palamau farmers is almost the same as that of the Hazaribagh farmers but in the HYV period the $R^2$ value has considerably declined for the pair wheat & gram for Palamau district. This may be due to the greater uncertainty and risk involved in growing wheat as compared to gram in the HYV period as measured by the coefficient of variation of gross return (see Table 6.9). This might also indicate the fact that the farmers could not substitute wheat for gram to the extent desired despite higher gross return for wheat. The overall homogeneity of regressions in the two sub-periods for the pair wheat & gram is rejected at 1 per cent level. In the case of Ranchi and Singhbhum districts we find that the farmers have substituted wheat for gram specially in the HYV period. The coefficients of the lagged dependent variable, yield-rate and gross return variables are significant in Ranchi whereas only the coefficient of the lagged dependent variable is significant in Singhbhum district.

$R^2$ in most of these regressions is not very high indicating that crop-allocation decisions of the farmers are influenced by some other factors of which, as noted earlier, irrigation is one important factor. To examine the importance of irrigation-facilities for crop-substitution during the rabi season, we regress the logarithm of the ratio of acreages in percentage on the total irrigation facilities made available to the farmers during the rabi season (as measured by the total irrigated area under major crops during rabi season) using equation C.4. Regression results are presented in Table 6.7. All the districts show positive and highly significant coefficients of the irrigation variable specially in the HYV period. The very fact that a larger variance is explained by the irrigation variable in the simple 2-variate regression equation (C.4) than by the other variables in the multiple regression equations (C.1 & C.2) indicates that it (irrigation) is a dominating factor in the crop-allocation decision making process of the farmers. We also find that the elasticity of substitution between

TABLE 6.6: **Showing regression results indicating the importance of pre-sowing rainfall on acreage substitution between maize and autumn rice grown during kharif season**

| District | Pairs of Crops | Period | Constant | Regression Coefficient | $R^2$ |
|---|---|---|---|---|---|
| Santhal Pargana | Maize / Autumn-rice | Pre-HYV | 6.260 | -0.031 (0.185) | 0.003 |
| | | HYV | 8.273 | -0.295** (0.099) | 0.527 |
| | | Combined | 8.018 | -0.282* (0.153) | 0.159 |
| Palamau | Maize / Autumn-rice | Pre-HYV | 5.587 | -0.021 (0.268) | 0.001 |
| | | HYV | 7.658 | -0.299* (0.133) | 0.387 |
| | | Combined | 8.093 | -0.397* (0.215) | 0.005 |
| Ranchi | Maize / Autumn-rice | Pre-HYV | 2.321 | -0.029 (0.154) | 0.005 |
| | | HYV | 5.039 | -0.408** (0.171) | 0.416 |
| | | Combined | 4.472 | -0.344 (0.244) | 0.099 |

Notes: (i) Regressions were fitted for other districts also but since the coefficients came out insignificant they were not presented.

(ii)Regression were also fitted for other pair of crops i.e. ragi and autumn rice but in this case the coefficient of pre-sowing rainfall came out insignificant in all the cases, though negative in most cases. One of the reason for the non-significance of the pre-sowing rainfall coefficient in this case may be the peculiar characteristics of ragi which (like maize) normally requires less water than autumn rice but it can also be grown under heavy rainfall conditions with good yield on slopy lands.

wheat and gram with respect to irrigation is considerably higher than the elasticity of subsitution with respect to gross retrun in all the cases. Though these (the irrigation coefficients) being absolute figures would not be strictly comparable, unless the relative return and irrigation variables are uncorrelated,[4] one would take them to

---

4. We observed (in the previous chapter) that the irrigation variable is highly associated with the other independent variables included in the acreage response equations like lagged area under wheat and gross return of wheat specially in the HYV period. We expect the same pattern of inter-correlations in this case also because most of the increased irrigation facilities have been used in growing wheat rather than gram or barley as noted earlier in Chapter 2.

indicate that the irrigation variable plays quite a dominating role in the crop-substitution process. It is also easy to understand why this should be so, for the returns in any year are determined by the irrigation water available for cultivation.

Thus we find a considerable change in the acreage in favour of wheat as compared to gram and barley in the HYV period in the districts of Chotanagpur region specially in Santhal Pargana and Ranchi where irrigation facilities during the rabi season were made available more readily and also in great amount as compared to other districts. It was noted that the farmers are responsive to the relative return rather than relative price alone. We observed that the elasticity of acreage substitution with respect to relative return is significant (in the combined period for the pair wheat & gram) only for the districts of Santhal Pargana and Ranchi which are relatively better endowed with the infrastructural facilities as compared to other districts of Chotanagpur region. This might indicate the fact that unless the basic infrastructural facilities are made available to the farmers they may not be motivated adequately to respond to the economic incentives and further they may not also be in a position to respond to them even if they are so motivated. Our findings in this regard are quite consistent with the recent observations made by the Agricultural Prices Commission, Government of India, New Delhi, as reported by Kahlon and Dhar [67].

## 6.6 CONCLUDING REMARKS

Our study in this chapter has revealed that the subsistence-farmers take relative return into account rather than relative price alone while substituting one crop for another. During kharif season, only marginal adjustment of area is reported in favour of maize as against autumn rice and ragi specially in the districts where maize is grown as a major crop as compared to autumn rice and ragi. In some cases, we find that the farmers are going on planting these crops in alternative years irrespective of price changes. This is, however, quite consistent with the practice of crop-rotation system followed by the farmers of Chotanagpur. In most cases the crop-substitution decision of the farmer is influenced by uncertain weather conditions and the availability of pre-sowing rainfall.

TABLE 6.7: **Showing regression results indicating the importance of irrigation facilities on acreage substitution between rabi crops**

$$\left( Equation\ fitted : \text{Log}\ \frac{A_{rt}}{A_{st}} \times 100 = a + b\ \text{Log}\ I_t \right)$$

| District | Pairs of Crops | Period | Constant | Regression Coefficient | $R^2$ |
|---|---|---|---|---|---|
| Santhal-Pargana | Wheat/Gram | Pre-HYV | 1.190 | 0.298*** (0.077) | 0.649 |
| | | HYV | 1.222 | 0.432*** (0.094) | 0.727 |
| | | Combined | -2.281 | 0.790*** (0.085) | 0.829 |
| | Wheat/Barley | Pre-HYV | 2.042 | 0.299*** (0.088) | 0.589 |
| | | HYV | -1.975 | 0.828*** (0.077) | 0.936 |
| | | Combined | -1.163 | 0.734*** (0.053) | 0.915 |
| Hazaribagh | Wheat/Gram | Pre-HYV | -0.440 | 0.637** (0.224) | 0.502 |
| | | HYV | 0.611 | 0.571*** (0.157) | 0.843 |
| | | Combined | -2.687 | 0.937*** (0.104) | 0.818 |
| | Wheat/Barley | Pre-HYV | 4.721 | 0.058 (0.372) | 0.003 |
| | | HYV | -0.747 | 0.760*** (0.063) | 0.927 |
| | | Combined | -0.874 | 0.769*** (0.100) | 0.765 |
| Palamau | Wheat/Gram | Pre-HYV | 0.482 | 0.387*** (0.085) | 0.722 |
| | | HYV | 2.806 | 0.181 (0.300) | 0.044 |
| | | Combined | -1.693 | 0.680*** (0.082) | 0.792 |
| | Wheat/Barley | Pre-HYV | 1.526 | 0.346* (0.155) | 0.383 |
| | | HYV | -0.436 | 0.576*** (0.116) | 0.756 |
| | | Combined | 0.939 | 0.422*** (0.044) | 0.835 |
| Ranchi | Wheat/Gram | Pre-HYV | 2.720 | 0.174 (0.102) | 0.268 |
| | | HYV | -0.126 | 0.579*** (0.034) | 0.973 |
| | | Combined | 1.465 | 0.394*** (0.030) | 0.903 |

| | | | | | |
|---|---|---|---|---|---|
| Singhbhum | Wheat / Gram | Pre-HYV | 3.008 | 0.027 (0.056) | 0.029 |
| | | HYV | -1.205 | 0.739*** (0.062) | 0.947 |
| | | Combined | -0.473 | 0.336*** (0.050) | 0.717 |

TABLE 6.8: **Showing mean ($\bar{X}$) and coefficient of variation (CV) of gross return of substitutable crops grown during kharif season and rabi season in the two sub-periods**

| District | $\bar{X}$ / CV | KHARIF SEASON | | | | | |
|---|---|---|---|---|---|---|---|
| | | Autumn rice | | Maize | | Ragi | |
| | | Pre-HYV | HYV | Pre-HYV | HYV | Pre-HYV | HYV |
| Santhal | $\bar{X}$ | 184.81 | 535.05 | 310.208 | 637.59 | 95.34 | 256.85 |
| Pargana | CV | 0.318 | 0.333 | 0.491 | 0.591 | 0.476 | 0.536 |
| Hazari- | $\bar{X}$ | 167.81 | 334.92 | 321.02 | 773.24 | 177.14 | 454.37 |
| bagh | CV | 0.335 | 0.411 | 0.476 | 0.445 | 0.466 | 0.716 |
| Palamau | $\bar{X}$ | 179.64 | 378.74 | 248.84 | 698.25 | 112.53 | 324.80 |
| | CV | 0.330 | 0.376 | 0.385 | 0.584 | 0.330 | 0.467 |
| Ranchi | $\bar{X}$ | 168.21 | 354.93 | 296.69 | 759.43 | 157.62 | 355.23 |
| | CV | 0.350 | 0.354 | 0.454 | 0.496 | 0.408 | 0.376 |
| Singhbhum | $\bar{X}$ | 191.13 | 429.79 | 332.14 | 1072.68 | - | - |
| | CV | 0.403 | 0.419 | 0.450 | 0.441 | - | - |
| Dhanbad | $\bar{X}$ | - | - | 319.99 | 875.28 | 188.05 | 407.39 |
| | CV | - | - | 0.453 | 0.418 | 0.495 | 0.316 |

| District | $\bar{X}$ / CV | RABI SEASON | | | | | |
|---|---|---|---|---|---|---|---|
| | | Wheat | | Gram | | Barley | |
| | | Pre-HYV | HYV | Pre-HYV | HYV | Pre-HYV | HYV |
| Santhal- | $\bar{X}$ | 348.18 | 1333.99 | 251.01 | 626.89 | 199.01 | 511.85 |
| Pargana | CV | 0.449 | 0.342 | 0.465 | 0.384 | 0.305 | 0.344 |
| Hazari- | $\bar{X}$ | 348.28 | 1155.02 | 222.46 | 671.17 | 226.24 | 540.60 |
| bagh | CV | 0.354 | 0.460 | 0.464 | 0.456 | 0.426 | 0.372 |
| Palamau | $\bar{X}$ | 308.02 | 1118.65 | 199.95 | 478.05 | 200.39 | 430.12 |
| | CV | 0.364 | 0.595 | 0.453 | 0.227 | 0.352 | 0.313 |
| Ranchi | $\bar{X}$ | 340.72 | 1377.90 | 221.70 | 525.49 | - | - |
| | CV | 0.432 | 0.482 | 0.531 | 0.258 | - | - |
| Singhbhum | $\bar{X}$ | 359.75 | 1359.80 | 231.75 | 679.99 | - | - |
| | CV | 0.371 | 0.544 | 0.579 | 0.376 | - | - |

Note: Average gross return (X) is given in rupees per hectare.

As regards the rabi crops, we find a considerable change in the area in favour of wheat as compared to gram and barley during the HYV period. Irrigation facilities provided to the farmers during the rabi season constitute the most important factor behind the decision of the farmers for substituting wheat as against gram and barley. As the indirect evidence suggests, the farmers are very much sensitive to the risk element in allocating the land among substitutable crops. The elasticity of substitution between wheat and gram with respect to relative returns ( in the combined period regression) is highest in Santhal Pargana district where the risk (as measured by the coefficient of variation of gross return) of growing wheat relative to gram and barley has considerably declined in the HYV period whereas Palamau farmers did not substitute wheat for gram to the extent desired despite a relatively higher gross return for wheat probably because the risk of growing wheat as compared to gram had considerably increased in the HYV period. On the whole non-price factors (like pre-sowing rainfall, irrigation facilities and tradition) appear to be more important than the price factor in influencing the crop-substitution decision of the farmers.

## APPENDIX VI

## ELASTICITY OF SUBSTITUTION BETWEEN THE AREA PLANTED UNDER THE CROPS

Let us suppose that an individual farmer grows only two crops. A simple model of decision-making regarding allocation of acreages between these two crops could be as follows. He is supposed to maximise

$$P'_1A_1 + P'_2A_2$$

$$\text{Subject to } f(A_1, A_2) \leq C_o \quad \ldots\ldots \quad (1)$$

Where $A_1$ and $A_2$ are the acreages under the two crops, $f(A_1, A_2)$ is the cost of growing crop 1 and crop 2 in $A_1$ and $A_2$, $C_o$ the total finance available to him for this purpose and $P'_1$ and $P'_2$ are the revenue per acre of the two crops. $P'_1$ and $P'_2$ are then to be taken to be given by

$$P'_r = P_r Y_r \quad (r = 1, 2, \ldots\ldots\ldots)$$

$P_r Y_r$ being the price and yield rate per acre of the r th crop respectively. The elasticity of substitution between the crops in response to changes in return per acre may be taken to be given, as in single product production-function analysis, by

$$\sigma_{12} = \frac{\partial(A_1/A_2)}{\partial(dA_1/dA_2)} \cdot \frac{dA_1/dA_2}{A_1/A_2} \quad \ldots\ldots\ldots(2)$$

and at the point of equilibrium the rate of product transformation

$$-\frac{dA_1}{d_{A2}} = \frac{f_1}{f_2} = \frac{P_2}{p_1}$$

Hence,

$$\sigma_{12} = \frac{\partial(A_1/A_2)}{\partial(P_2/P_1)} \cdot \frac{P_2/P_1}{A_1/A_2} \quad \ldots\ldots\ldots(3)$$

If $S_{12}$ is constant and equal to -b, b > ∩ then from (3) we have

$$A_1 / A_2 = a\,(P'_1 / P'_2)^b$$

or $\text{Log}\,(A_1 / A_2) = \text{Log}\,a + b\,\text{Log}\,(P'_1 / P'_2)$ ....... (4)

Hence the elasticity of substitution between $A_1$ and $A_2$ in response to

relative return changes may be estimated by fitting the relationship (4) to the data on acreages and gross returns [86 p. 141].

SEVEN

# OVERALL CROPPING PATTERN RESPONSE IN CHOTANAGPUR REGION-
## Inter District and Inter Temporal Variations

### 7.1 INTRODUCTION

In the previous chapter we have made a detailed analysis of the crop-substitution behaviour of Chotanagpur farmers, taking the case of individual pairs of competing crops grown during different cropping seasons. The pair-wise analysis, however, is not fully adequate to handle the overall changes in the cropping pattern which might result from inter-crop bilateral or multilateral substitution between the competing crops, in a comprehensive way. Even in the absence of inter-crop substitution in a particular season, the changes in the overall cropping pattern might take place due to changes in the seasonal cropping pattern. These may be due to the following reasons :

(a) The general price differentials of the crops grown during different cropping seasons might lead to changes in the seasonal cropping pattern, (b) Farmers' preference for growing early variety winter rice grown during kharif season for the purpose of sparing the land for growing rabi crops, like wheat, barley and gram, in the succeeding season, as most of the rabi crops are sown before the harvest of late variety winter rice, might lead to changes in the seasonal cropping pattern, (c) Due to frequent droughts and floods during the kharif season, the farmers might change their behavioural attitudes and take initiatives for increasing the area under rabi crops subject to the suitability of land and availability of other inputs, and (d) The utilisation of culturable waste land including fallow lands which abundantly exists in Chotanagpur region as noted earlier in Chapter 2, might also take place, if more irrigation facilities, credit facilities and chemical fertilisers etc. are made available to the farmers, as one would expect in the HYV period, resulting in changes in the seasonal cropping pattern over the period.

In this chapter we have made an attempt to estimate the changes in the overall cropping pattern and the factors underlying these changes over the period in the districts of Chotanagpur region.

## 7.2 INDEX OF OVERALL VARIATION IN THE CROPPING PATTERN

As an index of the overall variation in the cropping pattern we use the coefficient of variation (in percentage) of area planted under different crops (major cereals in this case viz. winter rice, autumn rice, maize, ragi, wheat and barley) in a particular year in the district. Symbolically the index is

$$I_{cpt} = CV\left(A_t\right) = 100\left[\frac{1}{m}\sum_{r=1}^{m}\left(A_{rt} - \bar{A}_t\right)^2\right]^{1/2} \Big/ \bar{A}_t$$

Where,

$I_{cpt}$ = Index of overall variation in the cropping pattern of a particular district in year t.

$A_{rt}$ = Area under r th crop in the year t (r = 1, 2 ...m and t = 1, 2, ...n).

$\bar{A}_t = \sum_{r=1}^{m} A_{rt} / m$ = Arithmetic mean of the area planted under rth crop in year t.

The above index measures the extent of variation in the area planted under different crops. It could also be taken as the index of crop-diversification (in the reverse sense). Any change in the index of overall variation in the cropping pattern should indicate a change in the cropping pattern as a whole but the converse may not always necessarily hold. The coefficient of variation (CV) has been used in this study because it provides a relative measure of variation and in this case provides a measure of changes in the area under different crops relative to the average planted area under the crops which to some extent takes the changes in the total planted area into account. It should be noted, however, that if the changes in area under the crop is proportional to the changes in the total area or if the changes for the different crops are mutually compensating, the coefficient of variation will not provide a representative index of actual change in the cropping pattern because a change in the CV may be observed without any change in the proportion of area allotted to the crops. Again our index does not take the substitutability characteristics of the crops into account. A better index can be prepared. Some of the alternative methods are listed in Appendix VII. However, the CV method has been found to be the most representative and comprehensive for our purpose and therefore has been used in this study despite its limitations. Further as noted earlier it provides an index of crop-diversification (in the reverse sense) which in itself along with its determinants should be of great interest. We later note how it behaves as a diversification index for different years and different districts in the two sub-periods.

We have prepared the index of overall variation in the cropping pattern for each district and for each year in the two sub-periods and presented them along with the mean and coefficient of variation of the index in Table 7.1. This table indicates that the diversification of crops has increased in the HYV period in all the districts (except Singhbhum) but the increase is not very high. This may be because of the characteristics of the mainly subsistence-type of agriculture, lack of irrigation facilities and risk aversion attitudes of the Chotanagpur farmers. Inter-district differences in the extent of crop-diversification is also indicated. Diversification is highest in Palamau and lowest in Santhal Pargana. The degree of diversification is approximately proportional to the area planted during rabi season, which is also easily understandable.

TABLE 7.1: **Showing the index of overall variation in the cropping pattern (i.e. index of crop-diversification in the reverse sense) with mean and coefficient of variation in the districts of Chotanagpur region in the two sub-periods**

Index of overall variation in the cropping pattern during the pre-HYV period in

| Year | Santhal Pargana district | Hazaribagh district | Palamau district | Ranchi district | Singhbhum district | Dhanbad district |
|---|---|---|---|---|---|---|
| 1956-57 | 187.8 | 161.1 | 100.7 | 126.2 | 140.7 | 158.2 |
| 1957-58 | 185.8 | 164.1 | 118.4 | 127.2 | 146.9 | 153.7 |
| 1958-59 | 179.3 | 160.6 | 98.3 | 123.9 | 145.9 | 154.0 |
| 1959-60 | 180.0 | 161.0 | 108.9 | 122.4 | 147.3 | 152.4 |
| 1960-61 | 180.5 | 165.3 | 96.4 | 125.5 | 145.5 | 153.9 |
| 1961-62 | 179.6 | 109.4 | 97.6 | 124.4 | 145.7 | 154.6 |
| 1962-63 | 177.8 | 162.9 | 101.8 | 123.6 | 151.2 | 154.2 |
| 1963-64 | 175.6 | 157.4 | 96.9 | 129.5 | 151.4 | 147.7 |
| 1964-65 | 182.2 | 163.2 | 100.4 | 117.7 | 152.8 | 152.3 |
| 1965-66 | 182.4 | 160.8 | 96.3 | 123.1 | 154.0 | 149.1 |
| Mean | 181.10 | 162.56 | 101.57 | 124.35 | 148.24 | 153.01 |
| Coefficient of variation | 0.0190 | 0.0191 | 0.0654 | 0.0241 | 0.0250 | 0.0182 |

Index of overall variation in the cropping pattern during the HYV period in

| Year | Santhal Pargana district | Hazaribagh district | Palamau district | Ranchi district | Singhbhum district | Dhanbad district |
|---|---|---|---|---|---|---|
| 1967-68 | 179.3 | 133.7 | 98.0 | 117.0 | 152.6 | 143.1 |
| 1968-69 | 190.1 | 155.1 | 102.7 | 116.8 | 151.3 | 146.4 |
| 1969-70 | 182.8 | 153.6 | 97.4 | 114.5 | 153.4 | 146.7 |
| 1970-71 | 182.7 | 145.5 | 94.9 | 117.2 | 152.0 | 144.3 |
| 1971-72 | 186.7 | 159.8 | 99.2 | 118.8 | 151.5 | 152.3 |
| 1972-73 | 179.6 | 147.5 | 92.2 | 118.6 | 151.5 | 142.9 |
| 1973-74 | 176.8 | 146.6 | 92.7 | 116.1 | 150.0 | 139.0 |
| 1974-75 | 171.9 | 148.9 | 87.0 | 113.1 | 150.0 | 144.3 |
| 1975-76 | 169.8 | 147.8 | 86.4 | 110.7 | 146.4 | 135.8 |
| 1976-77 | 168.4 | 141.1 | 92.3 | 108.4 | 149.2 | 128.2 |
| Mean | 178.81 | 147.96 | 94.28 | 115.12 | 150.79 | 142.40 |
| Coefficient of variation | 0.0381 | 0.0469 | 0.0525 | 0.0283 | 0.0126 | 0.0427 |

## 7.3 ALTERNATIVE SPECIFICATIONS OF THE REGRESSION EQUATIONS

One would expect that the traditional farmers, while taking decisions for changing the overall cropping pattern, might consider the overall cropping pattern in the previous year and try to adjust it in response to changes in the overall price pattern and other relevant factors. Therefore, we have regressed the coefficient of variation of area under different crops in the current year on the coefficient of variation of the farm harvest price in the previous year, coefficient of variation of area under different crops in the previous year and the irrigation facilities made available during the rabi season in the current year. Alternative specifications are given below :

D.1 $CV(A_t) = a + b\,CV(P_{t-1}) + u_t$

D.2 $CV(A_t) = a + b_1\,CV(P_{t-1}) + b_2 I_t + b_3\,CV(A_{t-1}) + u_t$

Where,

$CV(A_t)$ = Coefficient of variation of area under different crops (actual area in hectares) in year t, which is taken to be the index of overall variation in the cropping pattern.

$CV(P_{t-1})$ = Coefficient of variation of farm harvest prices (actual prices in Rs./Qtl.) of different crops in year t-1, which is taken to be the index of overall variation in the price pattern.

$I_t$ = Total irrigation facilities made available during the rabi season as measured by the total irrigated area under major crops during the rabi season in year t.

$U_t$ = Random error terms.

In equation D.1 we have introduced only the coefficient of variation of the farm harvest prices of rth crops in the previous period in an attempt to estimate the direct contribution of changes in the overall price pattern in explaining the year-to-year variation in the overall cropping pattern. In equation D.2 we have included two additional variables[1] viz. irrigation variable ($I_t$) and lagged dependent variable

---

1. We have also tried with annual total rainfall as an additional variable in equation D.2 and found positive coefficient for this variable in all the districts except Singhbhum but this was insignificant in all the cases therefore we have dropped it in the final selection.

[CV($A_{t-1}$)]. Lagged dependent variable is introduced as a traditional variable whereas irrigation variable is included to estimate the importance of irrigation facilities during rabi season in farmers' decision for changing the overall cropping pattern. One expects a negative coefficient for the irrigation variable because an increase in the irrigation facilities during rabi season may lead to an increase in the area under rabi crops which might result in a decline in the CV of area under the crops and hence in an increased crop-diversification.

## 7.4 REGRESSION RESULTS

Regression results are presented in Table 7.2. Regression equations are fitted for each district and for each sub-period and also for the combined period. We have also computed the F-ratios associated with the test for overall homogeneity of regressions in the two sub-periods and the results are given in Table 7.3. In all the districts $R^2$'s increase a great deal when irrigation is brought in. $R^2$'s have increased in the HYV period also. Irrigation coefficients as expected are negative, and in some cases significantly so, in all the districts. The price-pattern coefficients are insignificant except in a few cases. Coefficients of CV ($A_{t-1}$) are positive in majority of them. Except in Santhal Pargana and Palamau the pre-HYV period and HYV period response pattern differ. Details of the districtwise analysis in the two sub-periods are given below.

We find that the farmers of Santhal Pargana district are not significantly responsive to overall price pattern changes as indicated by the small insignificant but positive coefficient of the price pattern variable in the pre-HYV period. But in the HYV period irrigation facilities provided during the rabi season appear to be the most important factor influencing farmers' decisions. The coefficient of the irrigation varilable is highly significant with negative sign as expected. In the combined-period regression, the coefficients of both irrigation variable and lagged dependent variable are significant. However, the Chow test does not suggest any significant change in the overall cropping pattern behaviour of the farmers in the district between the two sub-periods.

**TABLE 7.2: Showing overall cropping pattern response equations estimated for the districts of Chotanagpur region in the two sub-periods and in the combined period**

| District | Period | Eqn. No. | Constant | Regression coefficients of independent variables | | | $R^2$ | $F_{(V_1, V_2)}$ |
|---|---|---|---|---|---|---|---|---|
| | | | | $CV(P_{t-1})$ | $I_t$ | $CV(A_{t-1})$ | | |
| 1 | 2 | 3 | 4 | 5 | 6 | 7 | 8 | 9 |
| Santhal-Pargana | Pre-HYV | D.1 | 175.04 | 0.212 (0.185) | .. | .. | 0.141 | 1.308 (1,8) |
| | | D.2 | 84.25 | 0.052 (0.208) | -0.0003 (0.0739) | 0.525 (0.349) | 0.399 | 1.325 (3,6) |
| | HYV | D.1 | 179.97 | -0.052 (0.035) | .. | .. | 0.003 | 0.022 (1,8) |
| | | D.2 | 248.20 | -0.389** (0.152) | -0.091*** (0.022) | -0.283 (0.315) | 0.874 | 13.87*** (3,6) |
| | Comb-ined | D.1 | 176.91 | 0.119 (0.178) | .. | .. | 0.024 | 0.444 (1,18) |
| | | D.2 | 104.51 | -0.137 (0.129) | -0.038** (0.014) | 0.450* (0.216) | 0.617 | 8.610*** (3,16) |
| Hazaribagh | | | 162.13 | 0.014 (0.206) | .. | .. | 0.001 | 0.005 (1,8) |
| | | | 202.27 | 0.045 (0.223) | -0.230 (0.170) | -0.211 (0.173) | 0.250 | 0.665 (3,6) |
| | | | 140.82 | 0.304 (0.627) | .. | .. | 0.042 | 0.153 (1,8) |
| | | | 240.95 | 0.406 (0.642) | -0.072 (0.093) | -0.658* (0.287) | 0.489 | 1.916 (3,6) |
| | | | 134.34 | 0.792** | .. | .. | 0.230 | 5.391** |

*Contd.* Tabfle 7.2

| 1 | 2 | 3 | 4 | 5 | 6 | 7 | 8 | 9 |
|---|---|---|---|---|---|---|---|---|
| | | | | (0.341) | | | | (1,18) |
| | | | 151.85 | 0.394 | -0.159 | -0.004 | 0.351 | 2.884* |
| | | | | (0.433) | (0.101) | (0.211) | | (3,16) |
| Palamau | | | 98.48 | 0.116 | .. | .. | 0.010 | 0.080 |
| | | | | (0.409) | | | | (1,8) |
| | | | 115.06 | 0.181 | -0.097 | -0.161 | 0.073 | 0.158 |
| | | | | (0.482) | (0.519) | (0.531) | | (3,6) |
| | | | 100.39 | -0.269 | .. | .. | 0.089 | 0.783 |
| | | | | (0.304) | | | | (1,8) |
| | | | 52.37 | -0.292 | -0.023 | 0.529 | 0.465 | 1.738 |
| | | | | (0.277) | (0.076) | (0.394) | | (3,6) |
| | | | 94.42 | 0.142 | .. | .. | 0.015 | 0.274 |
| | | | | (0.271) | | | | (1,18) |
| | | | 99.83 | -0.084 | -0.127** | 0.058 | 0.315 | 2.451 |
| | | | | (0.257) | (0.055) | (0.236) | | (3,16) |
| Ranchi | | | 133.66 | -0.339* | .. | .. | 0.642 | 14.37*** |
| | | | | (0.089) | | | | (1,8) |
| | | | 164.50 | -0.422*** | -0.625 | -0.243 | 0.807 | 8.38** |
| | | | | (0.097) | (0.477) | (0.194) | | (3,6) |
| | | | 113.55 | 0.074 | .. | .. | 0.041 | 0.346 |
| | | | | (0.126) | | | | (1,8) |
| | | | 92.84 | -0.007 | -0.036* | 0.210 | 0.765 | 6.517** |
| | | | | (0.075) | (0.015) | (0.295) | | (3,6) |
| | | | 116.77 | 0.122 | .. | .. | 0.036 | 0.672 |
| | | | | (0.149) | | | | (1,18) |
| | | | 59.92 | -0.083 | -0.047** | 0.525** | 0.653 | 10.033*** |
| | | | | (0.102) | (0.021) | (0.231) | | (3,16) |

*Contd. Table 7.2*

| 1 | 2 | 3 | 4 | 5 | 6 | 7 | 8 | 9 |
|---|---|---|---|---|---|---|---|---|
| Singhbhum | | | 140.55 | 0.371*** (0.098) | .. | .. | 0.640 | 14.21*** (1,8) |
| | | | 106.61 | 0.279*** (0.072) | 2.897** (1.139) | 0.227 (0.181) | 0.876 | 14.17*** (3,6) |
| | | | 149.82 | 0.046 (0.099) | .. | .. | 0.026 | 0.210 (1,8) |
| | | | 295.95 | 0.071 (0.075) | -0.152* (0.075) | -0.948 (0.767) | 0.627 | 3.364* (3,6) |
| | | | 144.17 | 0.254*** (0.085) | .. | .. | 0.333 | 8.99*** (1,18) |
| | | | 66.53 | 0.208*** (0.071) | 0.0002 (0.0290) | 0.526*** (0.161) | 0.600 | 8.014*** (3,16) |
| Dhanbad | | | 293.42 | 0.252** (0.076) | .. | .. | 0.581 | 11.08** (1,8) |
| | | | 248.82 | 0.338*** (0.081) | -1.711 (1.204) | -0.671* (0.344) | 0.744 | 5.816** (3,6) |
| | | | 137.19 | 0.176 (0.156) | .. | .. | 0.137 | 1.267 (1,8) |
| | | | 256.90 | -0.008 (0.096) | -0.541*** (0.136) | -0.748 (0.409) | 0.802 | 8.108** (3,6) |
| | | | 143.99 | 0.132 (0.150) | .. | .. | 0.041 | 0.770 (1,18) |
| | | | 84.58 | 0.041 (0.097) | -0.320** (0.140) | 0.430 (0.270) | 0.675 | 11.06*** (3,16) |

The Hazaribagh farmers are not very responsive either to the overall price pattern changes or to the irrigation facilities in both the periods, none of the coefficients being significants; $R^2$ is also low.

The coefficient of lagged dependent variable is negative and significant at 10 per cent level of significance in the HYV period, reflecting the crop-rotation-element noted in the earlier chapter also, in the farmers' decision. The overall homogeneity of regressions in the two sub-periods is, however, rejected at 1 per cent level of significance, but this may be due to the sudden shift in the overall cropping pattern in the HYV period as indicated by the average index of overall variation in the cropping pattern. The average index of overall variation in the cropping pattern has declined from 162.56 in the pre-HYV period to 147.96 in the HYV period, indicating greater diversification, the decline being the highest among the district (see Table 7.1). This may be due to the changes in the behavioural attitudes of the farmers.

In the case of Palamau district we find that all the coefficients come out insignificant in all the regressions except that of the irrigation variable in the combined period, which is significant at 5 per cent level of significance with proper sign. It should be noted that for the same equation we find very small $R^2$ values for this district indicating the fact that some other more important factors are operating behind the decisions of the farmers. Among the other factors, changes in behavioural attitudes of the farmers may be worth mentioning. The overall cropping pattern behaviour remains almost the same in the two sub-periods.

In the districts of Ranchi, Singhbhum and Dhanbad the price factor dominates over the non-price factors in the pre-HYV period but in the HYV period non-price factors(specially the irrigation factor) appear to be dominating over the price factor in explaining the year-to-year variation in the overall cropping pattern. It should be noted that the overall price pattern coefficient in the pre-HYV period is positive and significant in the districts of Singhbhum and Dhanbad but negative and significant for Ranchi district. This might be taken to indicate that a widening of the price-spectrum has led to crop-diversification in Ranchi and to crop-concentration specialisation in Singhbhum and Dhanbad and conversely in the pre-HYV period. This might also indicate that in Ranchi the farmers are indifferent towards price pattern changes. The overall homogenety of regressions in the two sub-periods is rejected at 1 per cent level of significance for Ranchi and Dhanbad and at 5 per cent level of significance for Singhbhum district.

## 7.5 CONCLUDING REMARKS

In this chapter we have examined the extent of changes in the overall cropping pattern and the factors underlying these changes over the period in the districts of Chotanagpur region. We have prepared the index of overall variation in the cropping pattern for each district and for each year in the two sub-periods. This index is also taken to measure the index of crop-diversification, we find that the diversification has increased in the HYV period in all the districts except in Singhbhum but the increase is not very marked. This may be due to various constraints facing the Chotanagpur farmers like poverty, lack of irrigation facilities, traditionality and the subsistence characteristics of their agriculture. The diversification is highest in Palamau district where the percentage of area under kharif crops is lowest and the percentage of total area sown irrigated is highest among the Chotanagpur districts.

The regression results reveal that except in a few cases the changes in the overall price pattern alone do not explain the year-to-year variation in the overall cropping pattern as indicated by very small $R^2$ values but the introduction of additional variables like irrigation and lagged dependent variable considerably increased the explanatory power of the regressions in all the districts and coefficients came out significant in most cases. It should be noted that even after the introduction of additional variables we find very small $R^2$ values for the districts of Palamau and Hazaribagh indicating that some other more important factors are operating behind the decision of the farmers. A plausible explanation is offered in terms of changes in the behavioural attitudes of the farmers. We find that the irrigation facilities provided to the farmers during the rabi season appear to be an important factor influencing farmers' decision for changing the overall cropping pattern in the Chotanagpur region specially in the HYV period and in those districts where manifold increases in the irrigation facilities have been observed. On the whole non-price factors dominate over the price factor in influencing the farmers' decision.

TABLE 7.3: **Showing computed F-ratios associated with the test of overall homogeneity of overall cropping pattern response equations in the two sub-periods**

(Equation number D.2)

| District | Computed F statistics with degrees of freedom in brackets i.e. $F_{(V_1, V_2)}$ |
|---|---|
| Santhal-Pargana | 2.387<br>(4,12) |
| Hazaribagh | 7.053***<br>(2,12) |
| Palamau | 0.622<br>(4,12) |
| Ranchi | 12.319***<br>(4,12) |
| Singhbhum | 5.307**<br>(4,12) |
| Dhanbad | 7.604***<br>(4,12) |

## APPENDIX VII

## ALTERNATIVE MEASURES OF OVERALL VARIATION IN THE CROPPING PATTERN

The coefficient of variation (CV) of area under different crops namely

$$I_{cpt} = \left[ \frac{1}{m} \sum_{r=1}^{m} \left( A_{rt} - \overline{A}_t \right)^2 / \overline{A}_t \right]^{1/2} \qquad \ldots\ldots(1)$$

Where $A_{rt}$ is the area under r th crop in the t th year
(r = 1, 2, ......m; t = 1, 2, .....n),

$\overline{A}_t = \frac{1}{m} \sum_{r=1}^{m} A_{rt}$ is the arithmetic mean of the area under the crops and I stands for the index of overall variation in the cropping pattern, has been used in this chapter as an index of overall variation in the cropping pattern.

Various other alternative measures can be used for preparing the index of overall variation in the cropping pattern. Instead of using the absolute values of area under the crops one can use the index number of area under different crops which will provide a measure of changes in the overall variation in the cropping pattern relative to a particular year (i.e. the base year). But since our purpose of preparing the index is to estimate year-to-year changes in the overall cropping pattern and the factors underlying these changes it would not make much difference whether absolute area or the index number of areas are used.

If the total area available for cultivation remains the same over the period then the simple measure of variance can be taken for this purpose i.e.

$$I_{cpt} = \frac{1}{m}\sum_{r=1}^{m}\left(A_{rt} - \overline{A}_t\right)^2 \quad \ldots\ldots\ldots(2)$$

If all the crops are substitutes, for each other, the following formulae may provide a better measure i.e.,

$$I_{cpt} = \frac{1}{m}\sum_{r=1}^{m}\sum_{r=1}^{m}\left(\frac{A_{rt}}{A_{st}}\right)^2 \quad \ldots\ldots\ldots(3)$$

'or'

$$I_{cpt} = \frac{1}{m}\sum_{r=1}^{m}\sum_{r=1}^{m}\left(\text{Log } A_{rt} - \text{Log } A_{st}\right)^2 \quad \ldots\ldots(4)$$

Where $A_{st}$ is the area under substitute (of r th crop) crop in year t.

Owing to the physical limitation of substitution possibilities between the crops and flexibility in the area under total cereals (due to the presence of culturable waste land and fallow lands) in the Chotanagpur region we find that the coefficient of variation method seems to be the most representative as well as convenient index for the purpose at hand and has, therefore, been used in this study.

EIGHT

# SUMMARY AND CONCLUSIONS

The study of supply responses in agriculture and their determinants on the one hand and constraints on the other becomes particularly important in the context of planning for economic development and the concomitant agricultural transformation and modernisation involved. Regions with special problems and backwardness problems naturally need special attention. Generally the farmers' responsiveness to price changes in a backward region is expected to be less marked than that in other advanced regions; the farmers in a backward economy usually operate under various constraints which are themselves the characteristics of backwardness. There is a need for objective analysis as to whether the farmers of a backward region are not willing to respond or their responsiveness is constrained by the handicaps under which they have to operate. The present study, focussing its attention on some of the aspects of agricultural supply responses in a tribal concentrated Chotanagpur region, is an attempt in this direction. It may be of some help to the planners in identifying regional problems in traditional agriculture as well as in ushering in modern agriculture with new agricultural practices.

In the present study we have considered the concept of supply response in a broader sense, and various aspects of aggregate supply response and individual crop supply responses for important crops have been examined in detail subject to the availability of data. Acreage substitution between pairs of competing crops have been examined in detail and the overall cropping pattern response to price pattern and other non-price factors are looked into.

The analysis is made for all the major cereals (which constitute the considerable proportion of gross area sown) individually as well as of the aggregate of all these crops and pairs of substitute crops in the two sub-periods (i.e. pre-HYV and HYV period) and combined period in the districts of Chotanagpur region. The differences in the supply responsiveness to price and non-price factors within and between the districts and in the two sub-periods are noted and causes underlying these differences examined. Our analysis covers the period from 1956-57 to 1976-77. The available data have been split up into pre-HYV and HYV period. Pre-HYV period relates to the years 1956-57 to 1965-66 whereas the HYV period corresponds to the years from 1967-68 to 1976-77.

Chotanagpur is a socially and economically backward state (Bihar) in India. Though the homogeneity of soil in the region is almost assured the districts within the region differ among themselves in many respects such as the magnitude of concentration of tribal population, infrastructural facilities, adoption of new agricultural practices, agricultural productivity, cropping pattern, urbanisation and industrial development etc. Agro-economic conditions required for growing the crops also differ between the crops within and between the cropping seasons. Chotanagpur farmers mainly grow cereals for their own consumption and also for their limited cash requirements. Farmers mainly depend on the monsoon rainfall for their crop production. Cropping pattern is biased in favour of kharif crops and the proportion of total area planted during the rabi season is very small because of the lack of secured irrigation facilities. The farmers have, however, been provided with some irrigation facilities during the rabi season in the HYV period. Farmers use the traditional method of production but recently they have started using new agricultural practices specially HYV seeds and chemical fertilisers in producing wheat. The poor farmers prefer the extensive farming techniques rather than intensive farming techniques as indicated by the growth rates of area under wheat being higher than the growth rates of productivity of wheat in the HYV period. This along with other considerations, provides us the ground for approximating output by area. Farm harvest prices (in absolute terms), in general, have increased over the period specially in the HYV period but there is no commensurate increase in the area under the crops except in the case of wheat (which constitute a very small proportion of total area sown) indicating that some other more important (non-price) factors are operating behind the decisions of the farmers. It has been noted, however, that the relative price of agricultural output has declined over the period specially in the HYV period implying that the terms of trade have gone against agriculture. The productivity and output of total cereals have remained almost the same or marginally declined but population has been monotonically increasing in the region over the period implying that the availability of output per capita has declined over the period resulting in increased poverty and stagnation in the economy.

Owing to the peculiar characteristics of the decision making units and peculiar production conditions under which the Chotanagpur farmers take decisions[1] we have hypothesized that the farmers may not be consciously formulating and following a precise quantitative decision-scheme but broadly similar surrogate considerations may be at work behind their decisions. Thus we have emphasized the informal decision-formulation-process actually followed rather than a formal one (which, of course, has its own value) and use a simple model for studying supply responses of Chotanagpur farmers which is based on the hypothesis of "a simple working rule" characterised by conservatism, traditionality, risk aversion, subdued responses to relatively persistent and relatively large changes. The hypothesis of the simple working rule leads to the heuristic formulation that area planted under the crop in the current period is a function of area planted in the previous period with adjustment made in response to price, yield rate of the crop in the previous period and other crucial relevant variables. An analytical model is developed which enables us to identify suitable shifter variables (such as pre-sowing rainfall and irrigation etc.) for incorporating them in our multiple regression analysis. It has been noted in this connection that our estimating equations are apparently similar to the Nerlovian partial adjustment model in its final derived form, the only difference lying in the generating process of the lagged dependent variable and the implications of its coefficients. In the Nerlovian framework it is the outcome of the adjustment mechanism assumed in the model whereas we have taken it as a traditional variable.

---

1. For detail discussion in support of this argument see section 4.4 in Chapter 4.

The parameters involved are estimated using the OLS estimate procedures. The log-linear form equation is preferred to the simple linear form because of its implications of multiplicative effect; besides, the coefficients give a direct measure of the elasticities involved. Error terms are assumed to be distributed normally, identically and independently with zero means. The standard Chow test has been used for testing the overall homogeneity of regressions in the two sub-periods. We have noted in this connection the asymptotic validity of these (t and F) tests due to the presence of lagged dependent variable as one of the regressors in the equations. We have used the Durbin-Watson test, as an approximate test, for testing the presence of serial correlation in the error terms in the case of equations estimated for the combined period with 20 observations and the null hypothesis of no serial correlation is accepted at 5 per cent level of significance in all the cases except in a few cases[2] for which the test becomes inconclusive.

Supply responsiveness of Chotanagpur farmers is indirectly examined in terms of acreage response to price and non-price factors with the help of estimated regression equations (related to different aspects of supply responses) in the light of the a priori hypotheses determined on the basis of the analysis (in Chapter 3) of the background profiles of the districts of Chotanagpur region in the two sub-periods. We have noted that all the factors influencing farmers' decision (which we feel relevant and included in the discussions at appropriate places) could not be included in our regression analysis due to the non-availability of relevant time-series data. Allowing sufficient degrees of freedom with the time series data available for analysis in each of the periods also acted as a constraint in including many regressors in the regression-equations. However, we have kept these factors, specially traditional crop-rotation practice and risk-aversion attitudes (risk is measured, in this study, by the coefficient of variation of gross return) of the farmers and their impacts, in mind, while interpreting the coefficients of the variables included in the regression.

Acreage response equations estimated for individual crops reveal

---

2. For detailed explanation see last paragraph in section 4.8 in Chapter 4.

that the price itself is not an important factor in explaining the variations in area under the crops. All the price coefficients are insignificant except in a few cases. However, the acreage elasticity to price is higher in the HYV period as compared to the pre-HYV period in many cases; the difference is not very large. Subsistence farmers appear to respond to gross-returns rather than prices alone specially in the case of wheat. The coefficients of the gross returns variable is positive and significant in most cases. The traditional factor (as represented by the lagged dependent variable), pre-sowing rainfall and irrigation factor are more important than price-considerations, as indicated by the significance of the coefficients of lagged dependent, pre-sowing rainfall and irrigation variables. The coefficients of lagged dependent variable in the regression for the Chotanagpur region as a whole estimated from the panel data are positively significant and do not significantly differ from 1 for all the crops and in both the periods except for wheat in the HYV period.[3] The coefficient of lagged dependent variable being equal to 1 may be taken to indicate that almost the same amount of land is traditionally maintained over the period. In some cases the coefficient of lagged-dependent variable come out negative indicating fluctuations and significantly so in the case of Ranchi district for winter rice.[4] The irrigation factor plays quite a dominating role in farmers' decision for planting wheat.[5] Overall homogeneity of the regression equations between the two sub-periods is indicated.[6]

---

3. It has already been noted that for wheat HYV seeds have been used and significant increasing trend in the area under the crop has been observed.

4. In the pre-HYV period it is not significantly different from -1 indicating regular oscillation whereas in the HYV period and combined period it lies between 0 and -1 indicating damped oscillation. It may also be taken to indicate the "traditional" propensity of the tribal farmers (tribal concentration is highest in Ranchi) to maintain a uniform amount of land under cultivation over the years, so that if for any reason less land is cultivated in any year, it is made up by bringing a larger amount of land in the next year and vice versa.

5. Detailed explanation in support of this argument is given in section 5.2.3. in Chapter 5.

6. Out of 37 F-ratios computed for this purpose (for all the six major cereals) only 5 came out significant (3 at 5 per cent and 2 at 10 per cent level of significance). Details are given in Table 5.7.

Acreage response is proved to be non-symmetric in the case of Palamau and Ranchi districts for winter rice and in the case of Hazaribagh, Dhanbad and Chotanagpur region as a whole (panel data) for wheat. The analysis of the overall acreage response pattern is almost the same as that of winter rice because of its dominance in the total area sown. The significantly negative relative price coefficient observed for Singhbhum district appears to be perverse but quite consistent with the socio-economic attitudes of the tribal (Ho) farmers concentrated in the district and indicates their backward sloping supply behaviour. Their backward sloping supply behaviour is also reflected in the non-symmetric acreage response equation estimated for winter rice. We observed negative coefficients for price and gross return variable in both the regimes (namely when price/return increases and when they fall) and significantly so in the later regime. A plausible explanation is offered in terms of the primarily fixed cash requirements of the farmers due to their limited taste and natural conservatism and consequently resulting in perverse response to price/return changes. Overall homogeneity of total acreage response equations between the two sub-periods is accepted in all the districts.

Crop-substitution analysis (in Chapter 6) again indicates the subsistence farmers' responsiveness to relative returns also rather than to relative price alone. During the kharif season only the marginal adjustment of area in favour of maize (for which HYV seeds are also available) as against autumn rice and ragi is indicated specially in the districts where maize is grown as a major crop as compared to autumn rice and ragi. Negative relative price coefficients specially in the case of Hazaribagh and Ranchi districts indicate the farmers' attitude of indifference towards the economic considerations in planting upland crops (namely maize, autumn rice and ragi) during the kharif season. This may, however, be due to the crop-rotation (which can also be taken to indicate a systematic crop-substitution) practised by the farmers of Chotanagpur specially in these two districts and might also indicate the fact that marginal price-differentials may not be sufficient to induce the farmers' to change their tradition-dictated and time-tested crop-rotation practice. Crop-substitution decisions of the farmers are also influenced by the uncertain weather conditions and the availability of pre-sowing rainfall during the kharif season. Irrigation facilities play quite a dominating role in the farmers' decision for substituting wheat for barley and gram in the HYV period. The elasticity of substitution between wheat and gram with respect to relative returns is significantly positive only in the case of Santhal Pargana and Ranchi districts which are relatively better endowed with the infrastructural facilities as compared to other districts of Chotanagpur region. This might also indicate that unless the basic infrastructural facilities are made available to the farmers they may not be motivated adequately to respond to the economic incentives and further they may not also be in a position to respond to them even if they are motivated. As the indirect evidence suggests, the farmers are sensitive to the differential risk-element in allocating the land among the substitute crops. On the whole non-price factors like pre-sowing rainfall, irrigation facilities and tradition appear to be more important than the price factor in influencing the crop-substitution decisions of the farmers.

The extent of changes in the overall cropping pattern[7] and the factors underlying these changes over the period is examined in Chapter 7. We observed that crop-diversification has increased in the HYV period in the majority of districts but the increase is not very large because of various constraints facing the Chotanagpur farmers. The diversification is highest in Palamau district whereas the proportion of gross sown area irrigated is highest among the districts of Chotanagpur region. Regression analysis reveals that except in a few cases the changes in the overall price pattern alone does not explain the year-to-year variation in the overall cropping pattern. Here again the irrigation facilities given to the farmers during the rabi season appear to constitute an important factor influencing farmers' decision for changing the overall cropping pattern in the Chotanagpur region especially in the HYV period and in those districts where manifold increases in the irrigation facilities have been observed.

Thus the analysis of the agricultural supply response in the Chotanagpur region, undertaken in the present study, reveals the

---

7. As an index of overall variation in the cropping pattern we have used the coefficient of variation of area planted under different crops in a particular year in the district. It could be taken as the index of crop-diversification (in the reverse sense).

almost inelastic nature of the supply responses to price changes. This would seem to apply not only in the pre-HYV period but also in the HYV period except in the case of wheat for which HYV technology has been used. In the case of wheat the responsiveness is revealed more in terms of changes in gross return rather than changes in the prices alone. Non-price factors like pre-sowing rainfall (during the kharif season) and irrgation facilities (during the rabi season) appear to be more important than the price factor in influencing farmers' decisions. An examination of the inter-district variations not only shows its consistency with district characteristics but also help in identifying the factors which intensify or restrain the degree of farmers' response to prices and non-price factors. Substantially higher responses in price/return are obtained from those districts in which the infrastructural facilities are relatively more abundant. The implication seems to be that the increased infrastructural facilities in the district induces the farmers to become more responsive to economic stimuli. Further, the variations in the elasticity estimates across the districts do not suggest that the socio-cultural attitudes of the tribal farmers act as a constraint on the agricultural supply response except in the case of Singhbhum where perverse price response is indicated but it is quite consistent within the present framework of our analysis.

All this leads to the conclusion that the low price responsiveness of the Chotanagpur farmers does not mean that they are not willing to respond but that they do not get an adequate opportunity to respond to price changes. Their price responsiveness is limited by infrastructural and other agricultural constraints. Unless these constraints are removed by the Government or other agencies through heavy investment and special care, the Chotanagpur farmers cannot fully be blamed for not being responsive to price and other stimuli. Apart from this main conclusion the study also indicates that for increasing agricultural output in a predominantly subsistence economy like Chotanagpur non-price stimuli would be of great help to the farmers. The Government should make other inputs, such as fertiliser, increased irrigation, pesticides and new strains of seeds suitable to the soils of Chotanagpur etc. available to the farmers. The price stimuli themselves in terms of relative agricultural price ratio - of agricultural price index to the overall wholesale price index - have not been very strong. Therefore, a boost of price incentive is also calles for. This, by itself, however, is not adequate. Provision of complementarities of increased agricultural output and agricultural development in adequate measure, and removal of infrastructural and other bottlenecks is urgently needed.[8]

---

8. Since our study is mainly based on the secondary sources of data, we could not examine the relationship of the individual characteristics of the farmers and their agricultural conditions in particular, the size of land holding etc. to agricultural supply responses. This requires individual level study through intensive field surveys. We could not conduct such surveys due to various financial and time limitations. However, we could conduct one survey of tribal village and the results are used to substantiate our arguments. Response studies at the individual farmer level for different group of size-holdings and for different income group for the two communities (tribal and non-tribal) would be helpful in establishing some more interesting propositions about the responsiveness of the Chotanagpur farmers and appropriate remedial measures could then be suggested for the overall development of the region.

# APPENDIX - A

## A NOTE ON SOURCES OF DATA AND SELECTION OF PERIOD

A major source for the data used here is the series of Annual Season and Crop Reports published by the Directorate of Statistics and Evaluation, Government of Bihar, Patna, providing district-wise area, yield rate and output data including information regarding land-utilisation, monthly rainfall and irrigation facilities. Farm harvest prices of individual crops (from 1954-55 to 1974-75) are taken from the Farm (Harvest) Prices of Principal Crops in India published by the Directorate of Economics and Statistics, Ministry of Agriculture and Irrigation, Government of India, New Delhi. Other sources include the office of the Directorate of Statistics and Evaluation, Government of Bihar, Patna, and the office of the Directorate of Agriculture (Agricultural Statistics Section), Government of Bihar, Patna, providing various relevant unpublished data. We have also benefited by various published and un-published reports (including village surveys) and even personal discussion with Agronomists, Social Scientists, Engineers and District Agricultural Officer of Ranchi and Hazaribagh district.

We now turn briefly the difficulties in using the major source of data, namely series of Annual Season and Crop Reports and in connection with the selection of period for the present study.

It was not found justified to use the pre-independence data, although they were available in season and Crop Reports from 1911-12 onwards because of its unreliability caused by the unscientific collection, on the one hand, and lack of the British Government's initiative on the other. Bihar like West Bengal and Orissa, having permanently settled estates with Zamindars and annual revenues fixed in perpetuity, the Government had no direct interest in keeping careful records of the area sown, crop grown, outturn and value of crops raised by individual cultivators, although such records have to be maintained in raiyatwari and temporarily settled areas. Agricultural statistics in permanently settled areas were prepared on the basis of records collected at the time of survey settlement operations usually held at the interval of 30 to 40 years, by making rough and ready adjustments in the light of weather conditions prevailing in the individual years.

Realising the necessity of reliable data for various development-schemes, the agricultural statistics section was set-up in the Revenue Department at the end of the 1944. The system of estimating acreage under various crops by the method of random sampling was first given a trial but it had to be given up in favour of a complete enumeration-survey with crop-cutting experiments on a sample basis, recommended by Professor P.C. Mahalanobis in 1945 (see, for detail, the article by Mahalanobis [78]). After four years of experiment, the reliability of the method was confirmed and the old method was completely given up and the work of issuing forecasts has been taken over by the Superintendent, Agricultural Statistics from the Director of Agriculture, Bihar since 1st April 1949 [50]. Reliable (relatively) data are available for paddy. wheat, gram and barley (the crops which have been included during the experimental period) only since 1950. It should be noted, however, that the traditional eye-estimation method for estimating output and yield rate in the case of maize and ragi under the study continued till it was replaced by the crop-cutting experiments method in 1959-60 and 1976-77 respectively [39].

Again, it is only since 1954-55 that separate yield-rates and output of maize and ragi are maintained districtwise. These are such important crops of this region that one cannot ignore them. It appears that before 1954-55, four crops viz., jowar, bajra, maize and ragi were combined in the name of "Indian Corn" as far as yield rate and output are concerned. However, acreage-data are available separately.

The geographical area of Chotanagpur districts (Singhbhum and Manbhum, the former Dhanbad) have been affected significantly by the readjustment of territories with Orissa and West Bengal in 1949 and 1956. In 1949 Bihar gained Sarai-Kela-Kharsawan sub-division from Orissa and the area of Singhbhum district was enlarged. "Under the implementation of the Bihar and the West Bengal Transfer of Territories Act 1956 entire Manbhum district of area 2407 square miles and population 1169097 excluding the police station of Chas, Chandankeary, Patamda, Chandil and Ichagarh together with a considerable portion of Purnea district has been transferred from Bihar to West Bengal. Consequently the Government of Bihar amalgamated the remaining police stations viz. Patamda, Chandil and Ichagarh into Singhbhum district and Chas and Chandankeary into the district of Dhanbad" [37].

Owing to the gradual development and frequent change in the authority of publication of the Annual Season and Crop Reports upto 1955-56, the early figures are not strictly comparable to the more recent ones, a fact which is clearly mentioned in every issue of the official publications relating to area and yield of crops. From 1911-12 to 1935-36 it was published by the Department of Agriculture, Bihar and Orissa. From 1936-37 to 1951-52 and 1952-53 to 1954-55, it was under the Revenue Department and Finance Department respectively of the Government of Bihar. In 1955 for the first time a separate independent body of the Government of Bihar was created for this purpose in the name of Central Bureau of Economics and Statistics also known as Directorate of Economics and Statistics. It was renamed as Directorate of Statistics in 1958 and then again changed to the Directorate of Statistics and Evaluation in 1960 which still continues.

Thus to maintain the consistency, comparability and reliability of the data, we finally decided to cover the period from 1956-57 to 1976-77 for the present study. The upper limit is restricted by the availability of the most recent data at the time of investigation.

# APPENDIX - B

## ADJUSTMENTS OF DATA

Adjustment is required to make the data consistent over the period, mainly due to changes in the unit of measurement in 1968 and creation of new districts in 1972. The adjustment of data also includes the estimation of some missing price figures.

The Annual Season and Crop Reports, the basic data source of the present study, maintained the yield rate and output of individual crops in maund (82.2857 lbs.) per acre and tons respectively upto 1967-68 and subsequently in kilogram per hectare and metric tonnes. Surprisingly, the presentation of area under the crops in acres is still continuing. Therefore area under the crops are converted into hectares (1 acre=0.4047 hectare) for the entire period and yield rates are converted in kilogram per hectare upto 1967-68. Output of the crop is the product of area under the crop and its yield rate. Gross return per hectare of the crop concerned is estimated by multiplying the yield rate with their respective farm harvest prices. Aggregate area under total cereals is computed by adding the area under the individual cereals included in the group of total cereals. Average yield rate and average price of total cereals are the weighted means of yield rates and farm harvest prices of the crops included in the group, weights being the area under the crop and output of the crop respectively. Gross return per hectare for the total cereals is computed by adding the gross return per hectare of the individual crops included in the group of total cereals.

Adjustment of data (for Hazaribagh district) was also needed due to the division of districts in 1972 which resulted in the creation of the new district "Giridih" within the same geographical area. Thus the relevant data for old Hazaribagh district are available only for six years in the HYV period (i.e) from 1967-68 to 1972-73) after that the data are available for two new districts (new Hazaribagh and Giridih) separately. Therefore, for the purpose of collecting more data (at least to the extent what we have for other districts i.e. upto 1976-77) for Hazaribagh district we have adjusted the data for the year from 1973-74 to 1976-77 for making consistent with the earlier ones. The following adjustments are made in this regard : (i) Area under the crop (sown, irrigated etc.) for the new districts of Hazaribagh and Giridih have been added to represent the area under the crop (sown, irrigated etc.) for old Hazaribagh district after 1972-73, (ii) Yield rate and price of individual crop for old Hazaribagh district are the weighted means of yield rates and prices of the crops in the two new districts, weight being the area under the crop and output of the crop respectively, and (iii) Monthly rainfall of the old district of Hazaribagh (after 1972-73) has been estimated by the weighted mean of the rainfall in the two new districts, the weights being the number of rain recording stations in two new districts.

It should be noted that a minor rearrangement of area between Dhanbad and Hazaribagh district during 1969-70, in which Dhanbad gained and Hazaribagh lost an equivalent area of about 43.53 square miles, remained unadjusted due to various technical difficulties.

Farm harvest price of autumn rice for Santhal Pargana district for the period from 1956-57 to 1958-59 is not reported. Estimated prices are used for these years. The prices are estimated on the basis of linear regression fitted, treating price of autumn rice in Santhal Pargana district as dependent variable and the price of autumn rice at state level as independent variable for the period 1959-60 to 1976-77. The regression relation between the price of autumn rice in the district of Santhal Pargana (Y) and the price for the Bihar State (X) comes out as

$$Y = 2.10 + 0.94\,X;\ R^2 = 0.91.$$

This relationship has been used to estimate the prices of autumn rice for Santhal Pargana district from 1956-57 to 1958-59 from the state average prices given for the respective years.

Price of autumn rice for the single year 1955-56 for all districts (for which state average price is also not available) and few other prices are estimated by assuming the 4th differences of the prices to be zero that is using interpolation based on the assumption of a cubic curve for the prices. For example, asssume the prices of t th year (t= 1, 2,.....5), in the relation Y = f(t), are $f(1) = p_1$, $f(2) = p_2$, $f(3) = p_3$, $f(4) = p_4$ (missing) and $f(5) = p_5$. We have to estimate the price of the 4th year in this case. Since we have only four prices we shall assume that f(t) is a polynomial of 3rd degree, as a consequence $D^4f(t)$ is to be considered as zero.

Now,

$$D^4f(1) = 0 \text{ or } (E-1)^4 = 0.$$

Where D & E are operators, so that

$$D = E-1.$$

It follows that

$$D^4f(1) = p_5 - 4p_4 + 6\,p_3 - 4\,p_2 + p_1 = 0.$$

Therefore,

$$p_4 = (p_5 + 6\,p_3 - 4p_2 + p_1)/4.$$

This method is used for estimating individual missing prices in the series using 4 years prices around the missing price. For detail illustration of the above method see Goon, Gupta and Dasgupta [51 p. 41].

# BIBLIOGRAPHY


1. Acharya, S.S. and Sengupta, J.K., "Acreage Substitution Between Jute and Rice", Arthniti, Vol.IX, No. 1&2, January-July 1966, pp.28-46.
2. Agha, N.A., "Extension Agency and Agricultural Development", in Waheeduddin Khan (ed.) Papers and Proceedings of the Workshop-Cum-Seminar on Rural Institutions and Agricultural Development, National Institute of Community Development, Hydrabad, 1972, pp. 330-337.
3. Ahmed, E., Bihar: Physical Economic Regional Geography, Ranchi University, Ranchi, 1965.
4. Askari, H. and Cummings, J.T., "Output Price Response in Agriculture: An Evaluation", Indian Journal of Agricultural Economics, Vol.XXIX, No.2, April-June 1974, pp.33-42.
5. _______, ,"Supply Response of Farmers with Heterogenous Land", Vol.XXXI, No.1, January-March 1976, pp.13-22.
6. _______, Agricultural Supply Response: A Survey of the Econometric Evidence, Praeger Publishers, Inc, New York, 1976.
7. _______, "Estimating Agricultural Supply Response with the Nerlovian Model : A Survey", International Economic Review, Vol.XVIII, No.2, June 1977, pp.257-292.
8. A,Van Exem, S.J., Basic Socio-Economic Attitudes of Chotanagpur Tribals, The Catholic Cooperative Society, Ranchi, 1973.

9. Bapna, S.L., Aggregate Supply Response of Crops : A Study in a Developing Region in India, Sultan Chand & Sons, New Delhi, 1980.

10. Barber, W.J., "Economic Rationality and Behaviour Patterns in an Under Developed Area : A Case Study of African Economic Behaviour in the Rhodesias", Economic Development and Cultural Change, Vol.8, 1960, pp. 237-251.

11. Batra, M.M., Agricultural Production : Price and Technology, Delhi School of Economics Monograph in Economics No.4, Allied Publishers Private Limited, New Delhi, 1978.

12. Becker, G., "Irrational Behaviour and Economic Theory", Journal of Political Economy, Vol.70, No.1, February 1962, pp. 1-13.

13. Behrman, J.R., Supply Response in Under Developed Agriculture : A Case Study of four Major Annual Crops in Thailand, 1937-1963, North Holland Publishing Company, Amsterdam, 1968.

14. Bell, C., "A Note on Perverse Producer Response to Changes in Prices", in David Lehmann (ed.) Agrarian Reform and Agrarian Reformism, Faber & Faber Ltd., 1974, pp.262-268.

15. Bell, P.F. and Tai, J., "Market Middlemen and Technology : Agricultural Supply Response in the Dualistic Economies of Southeast Asia", Malayan Economic Review, Vol.14, No.1, April 1969, pp.29-47.

16. Bhagat, L.N., "Adoption of New Agricultural Practices in Tribal Bihar : The View from a Village", The Economic Studies, Vol.22, No. 2, August 1981, pp.147-156.

17. _______, "Adoption of New Agricultural Practices in Bihar : A Study of Socio-Economic Implications", Unpublished Paper, 1981.

18. ________, "Land Concentration in Bihar : An Inter-Regional Analysis", The Economic Studies, Vol.22, No.11&12, 1982, pp.549-553.

19. _________, "Inter-Regional Disparities in Agricultural Infrastructure : A Case Study of Bihar", Indian Journal of Agricultural Economics, Vol. XXXVIII, No.1, January-March 1983, pp. 56-62.

20. ________, "New Technology and Agricultural Development in Tribal Areas : A Study of the Chotanagpur Region", in S.N. Mishra and Bhupinder Singh (eds.) Tribal Area Development, Society for the Study of Regional Disparities, New Delhi, 1983, pp. 3-20.
21. Bhagwati, J.N. and Chakrabarty, S., "Contributions to Indian Economic Analysis : A Survey, "The American Economic Review, Supplementary, Vol.59, September 1969, pp.2-73.
22. Bihar State Planning Board, Government of Bihar, A Draft Perspective Plan, Bihar 1978-89, Patna, 1978.
23. Bose, S.R., A Study in Bihar Agriculture, A.N. Sinha Institute of Social Studies, Patna, 1967.
24. _______,Economy of Bihar, Firma, K.L. Mukhopadhyay, Calcutta, 1971.
25 ._______,and Ghose, P.P., Agro-Economic Survey of Bihar, B.K. Enterprise (Publishers), Patna, 1976.
26. Cowling, K. and Gardner, T.W., "Analytical Models for Estimating Supply Relations in the Agricultural Sector : A Survey and Critique, "Journal of Agricultural Economics, Vol.15, No.3, June 1963, pp.439-451.
27. Cummings, J.T., "The Supply Responsiveness of Indian Farmers in Post Independence Period : Major Cereals and Cash Crops," Indian Journal of Agricultural Economics, Vol.XXX, No.1, January-March 1975, pp.25-40.
28. ________,"Crop Substitution and Market Influence: The Case of Wheat and Barley in North-Western India." Indian Journal of Agricultural Economics, Vol.XXXII, No.2, April-June 1977, pp.1-12.
29. Dasgupta,Biplab, Agrarian Change and the New Technology in India, United Nations Research Institute for Social Development, Jeneva, 1977.
30. Dantwala,M.L., "Incentives and Disincentives in Indian Agriculture," Indian Journal of Agricultural Economics, Vol.XXII, No.2, April-June 1967, pp.1-25
31. _______,"The Problems of a Subsistence Farm Economy: The Indian Case," in Clifton R. Wharton Jr.,(ed), Subsistence Agriculture and Economic Development, Frank Cass & Co. Ltd., London, 1970, pp.382-386.

32. _______,(ed.), "Symposium on Farmers' Response to Prices," Journal of the Indian Society of Agricultural Statistics, Vol.22, June 1970, pp.1-14.
33. Dean, Edwin, "The Supply Responses of African Farmers," North-Holand Publishing Company, Amsterdam, 1966.
34. Department of River Valley Projects, Government of Bihar, Bihar State Irrigation Commission Report 1971, Patna.
35. Deshpande, S.H., "Innumeracy and Economic Behaviour", Economic and Political Weekly, Vol.XVI, No.10, 11 & 12, March 1981, pp.501-513.
36. Directorate (Office of the) of Agriculture, Agricultural Statistics Section, Government of Bihar, Patna.
37. Directorate of Economics and Statistics, Government of Bihar, Hand Book of Statistics of Reorganised Bihar 1956, Patna.
38. Directorate of Economics and Statistics, Ministry of Food, Agriculture, Community Development and Cooperation, Government of India, Farm (Harvest) Prices of Principal Crops (1954-55 to 1974-75), New Delhi.
39. Directorate (Office of the) of Statistics and Evaluation, Government of Bihar, Patna.
40. Directorate of Statistics and Evaluation, Government of Bihar, Annual Season and Crop Reports 1955-56 to 1972-73.
41. _______, Bihar Trough Figures 1977, Patna, 1977.
42. _______, Bihar Statistical Hand Book 1978, Patna, 1978.
43. _______,"Study Report of the District-wise Economic Development in Bihar 1974-75" (in Hindi), Patna, 1978 (mimeo).
44. _______, "Study Report of Rural Debt and Investment Survey" (in Hindi), Patna, 1980 (mimeo).
45. _______, "Statement showing Wholesale Prices of Agricultural Commodities in the State of Bihar from 1961 to 1976", Patna (mimeo).
46. Durbin, J., "Testing for Serial Correlation in Least Squares Regression when some of the Regressors are Lagged Dependent Variables," Econometrica, Vol.38, No.3, May 1970, pp.410-421.

47. ______, and Watson, G.S., "Testing for Serial Correlation in Least Squares Regression", Vol.37, 1950, pp.409-428, Vol.38, 1951, pp.159-178, and Vol. 58,1971, pp.1-19.

48. Dutta, K., The Life and Thought of the People of Bihar, Scientific Book Agency, Calcutta, 1967.

49. Draft Sub-Plan for Tribal Region of Bihar 1974-79, prepared by Chotanagpur-Santhal Pargana Autonomous Development Authority, Ranchi, March 1974.

50. Ganguli, B.N., "The Economic Significance of Certain Statistical Concepts in Agricultural Economics", Sankhya, Vol.4, Part 2, 1938, pp.236-240.

51. Goon, A.M., Gupta, M.K. and Dasgupta, B., Fundamentals of Statistics, The World Press Private Limited, 4th Edition, Calcutta, 1971.

52. Goswami,P.C. and Saikia, P.D., "Problems of Agricultural Development-in Tribal Areas," Indian Journal of Agricultural Economics, Vol. XXV, No.3, July-September 1970, pp. 140-148.

53. Griffin, K., The Political Economy of Agrarian Change, The Macmillan Press Ltd., Second Edition, 1979.

54. Griliches, Z., "Hybrid Corn : An Exposition in the Economics of Technological Change," Econometrica, Vol.25, No.4, October 1957, pp.501-522.

55. Heady, E.O., "Uses and Concepts in Supply Analysis, in Earl O. Heady (ed.), Agricultural Supply Functions - Estimating Techniques and Interpretation, Iowa State University Press, Iowa, 1961, pp.3-25.

56. ______, and Dillon, J.L., Agricultural Production Functions, The Iowa University Press, 1961.

57. Herdt, R.W., "Disaggregate Approach to Aggregate Supply," American Journal of Agricultural Economics, Vol.52, No.4, November 1970, pp.512-520.

58. Houthakker, H.S. and Taylor, L.D., Consumer Demand in the United States 1929-1970, Analysis and Projections, Harvard University Press, Massachusetts, 1966.

59. Indian Council of Social Science Research, Alternatives in Agricultural Development, Allied Publishers Pvt. Ltd., New Delhi, 1980.

60. Ivern, F., Chotanagpur Survey, Indian Social Institute, New Delhi, 1969.

61. Jodha, N.S., "Intercropping in Traditional Farming Systems," Progress Report - 3, Economics Programme, International Crops Research Institute for the Semi-Arid Tropics ICRISAT Patancheru P.O. Andhra Pradesh, India, April 1979 (mimeo).

62. ______,"Role of Credit in Farmers' Adjustment Against Risk in Arid and Semi-Arid Tropical Areas of India, Economic and Political Weekly, Vol.XVI, No.42 & 43, October 17-24, 1981, pp.1696-1709.

63. John, P.V., Some Aspects of the Structure of Indian Economy, 1947-48 to 1961-62, Asia Publishing House, Bombay, 1968.

64. Johnson, D.G., "World Agriculture Commodity Policy and Price Variability," American Journal of Agricultural Economics, Vol.57, No.5, December 1975, pp.823-828.

65. ______, "International Prices and Trade in Reducing the Distortions of Incentives," in Theodore W. Schultz (ed.), Distortions of Agricultural Incentives, Indiana University Press, Bloomington, 1978.

66. Johnston, B.F., and Mellor, J.W., "The Role of Agriculture in Economic Development," American Economic Review, Vol.51, No.4, September 1961, pp.566-593.

67. Kahlon, A.S. and Dhar, T.N., "Approach of Agricultural Prices Commission to Agricultural Price Policy," Agricultural Situation in India, Vol.35, No.6, September 1980, pp. 457-459.

68. Kendall, M.G. and Stuart, A., The Advanced Theory of Statistics, Vol.2. Third Edition, Three Volume Edition, Carles Griffin & Company Limited, London, 1973.

69. Khatkhate, D.R., "Some Notes on the Real Effects of Foreign Surplus Disposal in Under Developed Economies" Quarterly Journal of Economics, Vol.76, No.2, May 1962, pp.186-196.

70. Krishna, Jai and Rao, M.S., "Dynamics of Acreage Allocation for Wheat in Uttar Pradesh : A Study in Supply Response", Indian Journal of Agricultural Economics, Vol. XXII, No.1, January-March 1967, pp.37-52.

71. Krishna, Raj, "Farm Supply Response in India - Pakistan : A Case Study of Punjab Region", Economic Journal, Vol. LXXIII, September 1963, pp.477-487.

72. _______, "Agricultural Price Policy and Economic Development," in Herman M. Southworth and Bruce F. Johnston (eds.), Agricultural Development and Economic Growth, Cornell University Press, Ithaca, New York, 1967, pp.497-540.

73. ______, and Raychaudhury, G.S., "Agricultural Price Policy in India : A Case Study of Rice," The Indian Economic Journal, Vol.28, No.3, January- March 1981, pp.16-34.

74. Ladejinsky, Wolf, "Green Revolution in Bihar, The Kosi Area : A Field Trip," Economic and Political Weekly, Review of Agriculture, Vol.IV, No.39, September 23, 1969, pp.A147-A162.

75. Lall, R.B., "Social Change Among Urban Oraon", Bulletin of the Bihar Tribal Research Institute, Vol.V, No.1, March 1963, pp.24-37.

76. Lavania, G.S. and Srivastava, G.C., "Production and Resource Use Pattern on Tribal and Non-Tribal Farmers in Hill Areas of Chotanagpur (Bihar)," paper read at the 38th Annual Conference of the Indian Society of Agricultural Economics and abstracted in the Conference Number of the Journal, Vol.XXXIII, No.4, October-December 1978, pp.53-54.

77. Madhavan, M.C., "Acreage Response of Indian Farmers : A Case Study of Tamil Nadu," Indian Journal of Agricultural Economics, Vol.XXVII, No.l, January - March 1972, pp.67-86.

78. Mahalanobis, P.C., "Report on the Bihar Crop Survey : Rabi Season (1943-44)," Sankhya, Vol.7, Part I, 1945, pp.29-106.

79. Maji, C.C., Jha, D. and Venkataramanan, L.S., "Dynamic Supply and Demand Models for Better Estimations and Projections : An Econometric Study for Major Foodgrains in the Punjab Region", Indian Journal of Agricultural Economics, Vol.XXVI, No.1, January-March 1971, pp.21-34.

80. Mathur, B.S. and Srivastava, L.L., "Increasing Productivity of Acid Soils in the Eastern Region - Bihar," in Symposium on Increasing Productivity of Acid Soils in the Eastern Region 1979, The Fertiliser Association of India, Eastern Region, Calcutta, January 1980, pp.89-100.

81. Mellor, J.W., "Increasing Agricultural Production in Early Stages of Economic Development : Relationships, Problems and Prospects," Indian Journal of Agricultural Economics, Vol.XVII, No.2, April-June, 1962, pp. 29-46.

82. ______, "The Functions of Agricultural Prices in Economic Development", Indian Journal of Agricultural Economics, Vol. XXIII, No.1, January-March, 1968, pp. 23-37.

83. ______, "The Subsistence Farmers in Traditional Economies", in Clifton R. Wharton Jr. (ed.), Subsistence Agriculture and Economic Development, Frank Cass & Co. Ltd., London, 1970, pp.209-227.

84. Mohsin, M.A., "Choice of Technology for Tribal Farmers," Journal of Social and Economic Studies, Vol. 6, No.1, 1978, pp.111-114.

85. ______ and Sinha, N.P., "Rainfed Agricultural Technology for Bihar," Paper presented at the Kharif Workshop of the Department of Agriculture, Government of Bihar, held at Patna on April 27-28, 1979 (mimeo.).

86. Mukerji, V. and Mukerji, K., "Substitution Between Superior and Inferior Food-Grains in India Between 1900-1958", in Ashok V. Bhuleshkar (ed.), Indian Economic Thought and Development, Popular Prakashan, Bombay, 1969, pp.137-156.

87. Narain, D., Impact of Price Movements on Area Under Selected Crops in India 1900-1939, Cambridge University Press, London, 1965.

88. ______, "Growth of Productivity in Indian Agriculture", Indian Journal of Agricultural Economics, Vol.XXXII, No.1, January-March 1977, pp.1-44

89. Narayan, N.S.S. and Parikh, K.S., "Agricultural Planning and Policy in Draft Sixth Plan : Will Farmers Fulfil Planner's Expectation?", Economic and Political Weekly, Vol.XIV, No.30, 31 & 32, Special Number, 1979, pp.1295-1306.

90. National Council of Applied Economic Research, Long-Term Projections of Demand for and Supply of Selected Agricultural Commodities - 1960-61 to 1975-76, New Delhi, 1962.

91. ______, Structure and Behaviour of Prices of Food Grains, New Delhi, May 1969.

92. Nerlove, Marc, The Dynamics of Supply : Estimation of Farmers' Response to Price, The Johns Hopkins Press, Baltimore, 1958.
93. ______, "The Dynamics of Supply : Retrospect and Prospect", American Journal of Agricultural Economics, Vol.61, No.5 (Proceedings Issue), December 1979, pp.875-888.
94. ______, Grether, D.M. and Carvelho, J.L., Analysis of Economic Time Series : A Synthesis, Academic Press, 1979.
95. Nowshirvani, Vahid, Agricultural Supply in India : Some Theoretical and Empirical Studies, Ph.D. Dissertation, Massachusetts Institute of Technology, 1968.
96. ______, "Land Allocation Under Uncertainty in Subsistence Agriculture", Oxford Economic Papers, Vol.23, No.3, November 1971, pp.445-455.
97. Operation Research Unit, Birsa Agricultural University, Operation Research Project Experiments, 1978-79, Ranchi, 1980.
98. ______, Operation Research Project Experiments, 1979-80, Ranchi, 1981.
99. Oury, Bernard, A Production Model for Wheat and Foodgrains in France (1946-1961), North-Holland Publishing Company, Amsterdam, 1966.
100. Pandey, M.S., The Historical Geography and Topography of Bihar, Motilal Banarasi Das, Bankipur, Patna, 1963.
101. Parikh, A., "Farm Supply Response : A Distributed Lag Analysis," Bulletin of the Oxford University Institute of Economics and Statistics, Vol.33, No.1, 1971, pp.57-72.
102. ______, "Market Responsiveness of Peasant Cultivators : Some Evidence from Pre-war India", Journal of Development Studies, Vol.8, No.2, January 1972, pp.291-306.
103. Prasad, K.N., The Economics of a Backward Region in a Backward Economy : A Case Study of Bihar in Relation to Other States of India, (2 Volumes) Scientific Book Agency, Calcutta, 1967.
104. Prasad, N., Land and People of Tribal Bihar, The Bihar Tribal Research Institute, Government of Bihar, Ranchi, 1961.

105. Prasad, P.H., "Alternative Strategy for Agricultural Development in Bihar," (Contributed), A.N. Sinha Institute of Social Studies, Patna, 1977.
106. Rao, C.H.H., Technological Change and Distribution of Gains in Indian Agriculture. Studies in Economic Growth, No.17, Institute of Economic Growth, The Macmillan Company of India Limited, New Delhi, 1975.
107. Rao, M.S. and Krishna, Jai, "Price Expectation and Acreage Response for Wheat in Uttar Pradesh," Indian Journal of Agricultural Economics, Vol. XX, No.1, January-March 1965, pp.20-25.
108. Rao, P. and Miller, R.L., Applied Econometrics, Prentice Hall of India Private Limited, New Delhi, 1972.
109. Ray, S.K., Cummings, B.W. and Herdt, R.W., "Agricultural Prices, Production and Growth," Economic and Political Weekly, Vol.XIV, No.39, September 29, 1979, pp.A97-A105.
110. Report of the Scheduled Area and Scheduled Tribe Commission, 1961, Government of India, New Delhi.
111. Report of the Study Team on Agricultural Credit Institutions in Bihar, Reserve Bank of India, Agricultural Credit Department, Bombay, 1977.
112. Revenue (Agricultural Census) Department, Government of Bihar, Report on Agricultural Census 1970-71, Patna, 1974.
113. ______, Agricultural Census (1976-77) : Number and Area of Operational Holdings, Patna, April 1979 (mimeo.).
114. Roy Burman, B.K. and Sharma, P.S., "Tribal Agriculture in India," Indian Journal of Agricultural Economics, Vol.XXV, No.3, July-September 1970, pp.149-160.
115. Ruttan, V.W., "Induced Innovation and Agricultural Development," Agricultural Development Council Staff Paper 77-1 reprinted from Food Policy, Vol.2, No.3, August 1977, pp.196-216.
116. Sachidananda, "Social Change in Chotanagpur", Bulletin of the Bihar Tribal Research Institute, Vol.VI, No.2, September 1964, pp.220-239.
117. Sawant, Shashikala, Supply Behaviour in Agriculture, Himalaya Publishing House, Bombay, 1978.

118. Sawhney, B.L., "Farm Supply Responses : A Case Study of the Bombay Region", Asian Economic Review, Vol. X, February 1968, pp.106-112.

119. Schultz, T.W., Transforming Traditional Agriculture, Studies in Comparative Economics No.3, Indian Edition by Lyall Book Depot, Ludhiana, 1970.

120. _______, "The Value of the Ability to Deal with Disequilibrium", Journal of Economic Literature, Vol.13, 1975, pp.827-846.

121. _______, (ed.), Distortions of Agricultural Incentives, Indiana University Press, Bloomington, 1978.

122. Sen, J., Community Development in Chotanagpur, The Asiatic Society, Calcutta, 1978.

123. Sengupta, J.K. and Sen, A., Econometric Supply Functions for Rice and Jute", Arthniti, Vol.12, No.1 & 2, January-July 1969, pp.1-40.

124. Shah, C.H., "Some Aspects of Agricultural Economics" in A Survey of Research in Economics, Vol.III, Agriculture (i), Indian Council of Social Science Research, New Delhi, 1975, pp.44-50.

125. _______ and Vakil, C.N. (eds.), Agricultural Development of India : Policy and Problems, Orient Longman Limited, New Delhi, 1979.

126. Shivamaggi, H.B., "Case Studies of a Few Agricultural Families Belonging to Weaker Sections in India", Reserve Bank Staff Occasional Papers, Vol.2, No.2, December 1977, pp.224-268.

127. Sidhu, D.S. and Kaul, J.L. "Composite Weather Index for Response Studies", Paper presented at the Inter-Disciplinary Symposium on Effects of Weather on Agricultural Production held on February 6-9, 1979 at Calcutta.

128. Sinha, S.P. and Varma, B.N., "A Study of Trends and Variations in the Prices of Food Grains in Bihar with Special Reference to Prices of Cereals Between 1956 and 1968", paper read at 31st Annual Conference of the Indian Society of Agricultural Economics and abstracted in the Conference Number of the Journal,- Vol.XXVI, No.4, October-December 1971, pp.446-447.

129. Singh, G., "Economics of Human Capital of Tribal Bihar," paper read at the 62nd Annual Conference of the Indian Economic Association and abstracted in the Conference Number of the Journal, Vol.1, Part 2, December 1979, p.106.
130. Singh, K., A Study of Impact of Green Revolution on Different Classes of Farmers (Thakurgaon Village in Ranchi District), Unpublished M.Sc. (Agriculture) Thesis, Ranchi University, 1974.
131 Singh, K.P. "Ratan", Farmers' Response Towards HYV of Paddy, Unpublished M.Sc. (Agriculture) Thesis, Ranchi University, 1977.
132. Singh, R.P. and Kumar, A., Monograph of Bihar : A Geographical Study, Bharti Bhawan, Patna, 1970.
133. Singh, T.B., Response of Dwarf Wheat (S-227 Kalyan Sona) to Different Levels and Forms of Nitrogenous Fertiliser, Unpublished M.Sc. (Agriculture) Thesis, Ranchi University, 1969.
134. Subbarao, K., "Farm Supply Response: A Case Study of Sugarcane in Andhra Pradesh," Indian Journal of Agricultural Economics, Vol.XXIV, No.1, January-March 1969, pp. 84-88.
135. Thamarajakshi, R., "Role of Price Incentives in Stimulating Agricultural Production in a Developing Economy", in Douglas Ensminger (ed.), Food Enough or Starvation of Millions, Tata McGraw - Hill, New Delhi, 1977.
136. The Fertiliser Association of India, Soils of India, New Delhi, 1972.
137. ______, Fertiliser Statistics 1977-78, New Delhi, 1978.
138. Tyagi, D.S., Farmers Response to Agricultural Prices in India, Heritage Publishers, Delhi, 1974.
139. Vidyarthi, L.P., Cultural Contours of Tribal Bihar, Punthi Pustak, Calcutta, 1964.
140. ______, Socio-Cultural Implications of Industrialisation in India : A Case Study of Tribal Bihar, Ranchi University, Ranchi, 1971.
141. Wharton, C.R. Jr., "The Economic Meaning of Subsistence", The Malayan Economic Review, Vol.8, No.2, October 1963, pp.46-58.

142. ______, "The Infrastructure for Agricultural Growth", in Herman M. Southworth and Bruce P. Johnston (eds.), Agricultural Development and Economic Growth, Cornel University Press, Ithaca, New York, 1967, pp.107-142.

143. Wipf, L.S. and Bawden, D.L., "Reliability of Supply Functions Derived from Production Function Estimates", American Journal of Agricultural Economics, Vol.51, No.1, February 1969, pp.170-178.

144. Xavier Institute of Social Service, Report of the Bench Mark Survey of Eleven Villages in Kanke and Ormanjhi Block of Ranchi District in Bihar 1977-78, Ranchi, 1978 (mimeo.).

145. ______,Action Proposal for Rural Reconstruction Based on Potential Resources of Five Villages in Chandil Block (Singhbhum District) to TELCO, Ranchi, March 1979 (mimeo.).

# Index